TECHNOLOGY: A SECOND LEVEL COURSE

# DESIGN
## PRINCIPLES AND PRACTICE

BLOCK 6

# A REVIEW OF DESIGN

PREPARED FOR THE COURSE TEAM BY
ED RHODES
(SECTION 6 WITH ROBIN ROY
AND STEPHEN POTTER)

## T264 DESIGN: PRINCIPLES AND PRACTICE

## CONTENTS OF THE COURSE




The Open University, Walton Hall, Milton Keynes, MK7 6AA.

First published 1992.





Edited, designed and typeset by the Open University.

Printed in the United Kingdom by The Burlington Press Cambridge Limited.

This text forms part of an Open University Second Level Course. If you would like a copy of *Studying with the Open University*, please write to the Central Enquiry Service, PO Box 200, The Open University, Walton Hall, Milton Keynes, MK7 6YZ, United Kingdom.
If you have not already enrolled on the Course and would like to buy this or other Open University material, please write to Open University Educational Enterprises Ltd, 12 Cofferidge Close, Stony Stratford, Milton Keynes, MK11 1BY, United Kingdom.

ISBN 0 7492 6145 5

Edition 1.1

# CONTENTS

# STUDY GUIDE

## INTRODUCTION

Earlier parts of the course have emphasised the centrality of user needs as a factor shaping approaches to design. Obviously, as a potential user, you are likely to be thinking carefully about your use of this Block. So, the question we have to try to answer at the start of every Block – What's in it for you as the student, or as 'the consumer'? – is particularly important here.

In the absence of market research, I have sought to meet what I believe are most likely to be your immediate needs in a number of ways. At this stage of the course, the Examination is likely to be predominant in your concerns. So the emphasis on 'review' in the Block title and content is a deliberate attempt to assist your revision by reminding you of some of the key concepts and issues which have been raised in Blocks 1 to 5. However, this does not mean that I am simply trying to re-serve 'cold portions' of earlier material, nor that this is a passive process. I have sought to avoid this in two ways.

One of these ways is through a fairly extensive provision of activities, including two fairly lengthy exercises in Sections 1 and 3. These activities will involve you in thinking back to, and applying, some of the ideas and skills introduced earlier in the course in an investigation and assessment of design issues and problems. One of the differences is that, rather than focusing on a single product, these activities will focus your attention on a group of products – textiles of various types. This product category has the advantage of being readily accessible for you to examine, and the variety of products it contains raise some useful questions. One set of these questions concerns the relationships *between* products in both product design and in the development of new types of product. But some textile products may appear to be significant in design terms only at the relatively superficial level of 'styling'. Furthermore, many textile products do not obviously offer the technological complexity and diversity which were evident, for instance, in Block 5's automobile study. Yet, as you will see, taken together, textile products present a surprisingly complex range of design issues. Investigating these issues should prompt you to reach back for some of the ideas and concepts introduced in previous Blocks, and you will find that there are some useful comparisons between textiles and other types of product.

But the investigation of textile products in this Block is not solely concerned with reviewing earlier material, and has another, distinct, purpose. Earlier parts of the course have referred to the way in which design generally has to be undertaken within companies or by those acting as design consultants to companies. These companies have a wider range of functions – as manufacturers and so on. In developing product strategies, in commissioning specific designs, and in other ways, they interact with a variety of external pressures and factors. For example, the development of a specific design may be shaped by safety and other legislation, by product market conditions, and by other companies such as retailers and materials suppliers. These pressures can impinge directly on the process of developing a design, and part of the skill of a designer comes from learning how to work within or around them. Thus, this Block has a more specific concern – that of considering how these wider organisational, economic and other factors can affect design, and how they can be responded to.

This task will partly be accomplished through a two-part 'Textiles Exercise' which examines how the apparent characteristics of clothing and other product designs are linked to particular competitive strategies, demand conditions and so on. From this groundwork, I go on in Section 5 to develop some more generally applicable conclusions about the significance of these 'contextual' factors for design practice. But, both in the Textiles Exercise and in Section 5, I am also concerned to start looking forward to the ways in which design practice may develop in the medium-term future. For example, it is important to think about the way in which product design is changing as the impact of CAD systems becomes more extensive. As you saw in Block 5, changes in the areas directly concerned with design are generally accompanied by similar changes in other areas of manufacturing concerns, and this is altering the way that product design is linked to the manufacturing process. In part, these changes follow from increasing links between electronically-based systems, but these links also increasingly extend outside the individual factory and company to provide direct links with other firms in the 'supply chain' – their suppliers and customers. Thus, a further important question is how product design may be changed as organisations increasingly take advantage of electronically-based integration to make direct inputs into the various stages of product design across a supply chain.

The theme of 'looking forward' to the evolution of design practice is taken into a specific area of concern in Section 6. It has been pointed out in earlier Blocks that, behind recent developments in production technologies, in product markets and so on, there are increasing uncertainties about the impact on the environment of our use of specific products and materials. To these concerns are added others arising from advancing levels of industrialisation and consumption across the world. It seems unavoidable that approaches to product design will have to undergo some fundamental changes in order to conserve scarce resources and to reduce the effects of human economic activity on the global ecological system. This will be a long-term process, and its probable characteristics are still not clear. But, given a need for action of some sort, I will ask you to consider what sorts of changes in design practice may be necessary.

## AIMS

As indicated by the Introduction, the aims for this Block are:

- to provide a recapitulating and integrative component for the course;
- to apply skills and experience derived from earlier parts of the course in an investigation of design issues presented by a broad category of domestic products, namely, textiles;
- to identify salient changes in the technological, economic and organisational contexts affecting the design of products of different types;
- to consider the changing tasks and responsibilities of designers in the context of technological, economic and organisational change;
- to review the possible implications for product design of environmental problems.

## OUTCOMES

At the end of your study of this Block, you should be able to do the following:

- summarise some of the main themes that run through the whole course;
- transfer the skills and awareness gained from previous Blocks of the course to a new product area;
- outline ways in which changes in technical, economic and organizational contexts can affect product designs;
- outline ways in which the role of the product designer may change under the influence of contextual factors;
- discuss the influence of environmental and global ecological concerns on product design.

**A detailed Checklist of Objectives is given near the end of the Block.**

## WHAT YOU HAVE TO DO

Block 6 of this course comprises this main Block text, with its exercises and self-assessment questions.

The Study Chart shows the suggested pattern of study for the Block. Directions are included at appropriate points within the Block for carrying out the two parts of the Textiles Exercise, for the Green Design Exercises, and for other activities accompanying the text. The Textiles and Green Design Exercises (printed in brown type) are fairly lengthy – the two parts of the Textiles Exercise will each require about an hour, and the two parts of the Green Design Exercise will each take about twenty minutes.

You should arrange your work on the Block to allow for these exercises to be completed before you read the subsequent sections. This is particularly important in the case of the Textiles Exercises because these will need to be undertaken in your home (or other place of residence). In carrying out the exercises, and in your work on subsequent sections, you should remember that the purpose of the Block is not to teach you how to undertake textile design or green design as such, but to remind you of, and to further develop your understanding of, more general design principles and important issues of practice as they are evolving.

The Block has been written with revision specifically in mind. You will be regularly referred back to concepts and ideas from earlier parts of the course. You should use these references back to check your own recollection and understanding of these and other, related points. Thus, you are likely to find it helpful to have the other Blocks near to hand, and to look back through them as appropriate. The video, audio and other components should be used similarly where this is possible.

Additionally, since you will regularly be thinking back across the course as a whole, you may also find it helpful to look through the Specimen Examination Paper prior to studying the Block and again at the end, and to think about how your more general revision for the Examination should be planned. You should also listen to the audio-cassette, which gives you advice on preparing for the Examination.

# STUDY CHART

| Section | Main text | Exercises | Video | Audio |
|---|---|---|---|---|
| 1 | An integrating exercise: Textiles Exercise, Part 1 | Textiles Exercise, Part 1 | | |
| 2 | The products and their characteristics | | | |
| 3 | Textiles Exercise, Part 2 | Textiles Exercise, Part 2 | | |
| 4 | A design chain? | Tailoring to shape<br>A fabric clamp | | |
| 5 | Design and organisation | Organisational influences | | |
| 6 | 'Green design' | Thinking about green design<br>Specimen Examination Paper | | Audio 4, Side 2, Exam preparation |

# 1 AN INTEGRATING EXERCISE: TEXTILES EXERCISE, PART 1

## 1.1 INTRODUCTION

The design of products in the textile sector – from fibre production through to the manufacture of products such as clothes for sale in retail outlets – presents a number of novel and complex design issues. Not least, these derive from the large overall size of the market. In the 1980s clothing accounted for about 7% of total household expenditure in the U.K., an expenditure of about £17 billion annually (compared with around £16 billion spent each year on the purchase of new automobiles). When household textiles and other types of textile product are included, textiles as a whole account for about 10% of direct household expenditure. As we will see, textiles are also 'consumed' in other ways, through their use in other sectors, for instance, in the health service and in most industrial sectors such as food manufacture, paper-making, agriculture, and building and construction.

The accessibility of textile products – they are all around us – and their seeming simplicity provide a useful focal point for applying some of the understanding of design which you have built up through the course. Examining these products also provides a means for looking again at some of the core issues and questions raised in the course as a whole, and examining the factors which are important in their design emphasises the general applicability of the material in the earlier parts of the course to consumer products in general. Textiles also provide some helpful contrasts in perspective. For instance, what are the implications when – as is sometimes the case – retailers as well as manufacturing companies employ design teams? Additionally, as I will show in later sections, patterns of technological and other change across the textile industries as a whole raise some fundamental questions about current assumptions in, and approaches to, product design.

Initially, you may feel somewhat sceptical about this assertion of complexity. One reason for scepticism is, perhaps, the association of many textiles with the home. Not only are textile products found mainly around the home, but they were for a long time *made* in the home. In Britain, as in the rest of the world, home production of knitted, sewn and other products still remains significant, both for personal use and as a money-earning activity. For instance, a 1980s study of employment in London indicated that 44% of workers in the women's dress industry were homeworkers (Leigh *et al.*, 1984). The fact that much production and design work is undertaken by women also runs up against the wider problems of gender dynamics in the workforce, and the associated under-estimation of the skills traditionally exercised by women.

But a further factor is that textile products have been around for a long time, and many of them appear very simple. Many are produced by methods which, in their essentials, have evolved from technological practice and experience which is of very long standing indeed. The earliest known textiles – linen from Catal Hüyük in Anatolia – date from about 6000BC, while the use of sewing to assemble garments made from animal skins appears to go back much further – to the Palaeolithic era, about 15,000 years ago. Finds from later periods, such as the Iron Age clothing in Figure 1, demonstrate early knowledge of a range of animal and plant fibres which could be used to make textile artefacts for a variety of purposes, whether as clothing of different types and grades or as other textile products for domestic purposes and in work outside the

**FIGURE 1**
IRON AGE COSTUME RECOVERED FROM A NORTH EUROPEAN PEAT-BOG. THE SKIRT AND SCARF ARE OF WOOL, IN A TWILL WEAVE OF A CHECK PATTERN. THE CAPE IS OF SHEEPSKIN

**FIGURE 2**
MRS MARGARET BRIDGEMAN OF LLANGYNWYD, MID GLAMORGAN. PHOTO TAKEN BETWEEN 1906 AND 1914

home – in farming and fishing for instance. These early technological skills included the ability to spin yarns, to form fabrics by weaving and other means, and to process them, such as by using plant and other dyes to produce colour and pattern (Wild, 1988).

In one sense then, textiles might be regarded as a somewhat curious anachronism – a category of 'pre-industrial products' which have somehow survived into the modern age. Technological change in other sectors has resulted in the development of radically new products – from bicycles to videos – which had not existed before. We continue to use knitted or woven fabrics for clothing and other purposes, and to assemble them, in the main, by sewing – the sewing machine being 'no more than a power-operated needle' (Carr & Latham, 1988). This sense of a pre-industrial carry-over is reinforced by the continued survival of thriving hand-craft sectors in the industrialised as well as the industrialising countries. Even the pioneering role of the mechanisation of spinning and weaving in the first Industrial Revolution of the mid eighteenth century might be viewed as contributing to this sense of anachronism, not least because of the lingering image of 'satanic mills'.

But continuities with the far distant past need to be set against some fundamental discontinuities. One of these derives from a transformation in the various bodies of technological knowledge which provide the foundation for design (in the comprehensive sense in which you have come to understand 'design' through the course) across the various sectors involved in textile production. As you will see in Section 4, the knowledge base has become highly complex, and practice in both design and manufacture is changing rapidly. Behind this change, there is the comparatively recent emergence of a mass consumer market.

Up to the early twentieth century, textile products were, for most people in Britain, expensive basic necessities, made to last as long as possible – by repairs and by recycling clothes as 'hand-me-downs', through the large second-hand market, and by recycling discarded clothing by 'dismantling' it (reducing it to the basic wool or cotton fibres for purposes such as re-manufacture in 'mungo' or 'shoddy' fabric for low-cost clothing). By and large, fashion in clothing, as in other respects, was a phenomenon confined to the small upper and middle classes. For the rest of the population, functionality remained paramount – a factor in the surprising similarities between the clothing of Mrs Bridgeman in Figure 2 and that of some two to three thousand years earlier which is shown in Figure 1. The emergence of a mass market for ready-made products akin to that of today came only after 1914:

> Until World War 1 the women of the working class [in Britain] were still confined to the nineteenth century costume of a serge or tweed skirt, a blouse of sateen or velvet, flannel petticoats and woollen or cotton stockings. Women's heavy clothing industry, producing coats and skirts, had been organised on a factory basis in the second half of the nineteenth century, but in the light clothing industry, no factory organisation existed before World War 1. Women of the working class continued to buy lengths of dress material which they either made up themselves or had them made up by a dress maker, or patronised second-hand shops dealing in women's cast off clothing. But [during and] after the war, an increasingly large number of the working-class women and girls were employed in workshops and factories and there they discovered that their clothes were totally unsuited for work. They were heavy, unhealthy and inconvenient, they constricted movement, and the factory conditions could be hazardous and dangerous. At the same time, with their new-found independence, they began to feel it a mark of inferiority to wear the soiled and cast-off clothes of other women, while factory employment did not give them either the time or the inclination to make clothes for themselves. The rising wave of emancipation of women was marked by a growing demand for clothes that were inexpensive, light, comfortable and attractive, as befitted women's new status in society. [One firm in particular – Marks and Spencer – was] largely responsible for accelerating the development process of [the] light-clothing industry for women. An entirely new range of products were 'created' to satisfy the new mass market for cheap, well made and pretty clothes for working class women and girls at a price within reach of their household income.
>
> (Tse, 1985, pp.28–29)

**FIGURE 3**
UTILITY CLOTHING, EARLY 1940S

Parallel developments in the design, production and retailing of men's clothes contributed to a general – though not universal – increase in clothing purchases in the 1920s and 30s. In Britain, as in other countries, this change, and the extension of the fashion market to a wider section of the population, constituted the 'spearhead of consumerism' (Wilson & Taylor, 1989) – a precursor of the more general emergence of mass consumption. This process was interrupted by the shortages of the Second World War, during which clothing became regulated by wartime regulations and effective designs were fostered under the Utility Scheme in the same way as was discussed in Block 1. The results, illustrated in Figure 3, were similar in terms of encouraging economical, simple and elegant designs, and more generally in fostering improved approaches to design and to manufacture.

In the subsequent period, the 'spearhead' of 1930s clothes designing and retailing evolved into the rapidly changing, highly diverse, fashion-driven sector we are familiar with today. People now own a far more diverse range of textile products than their counterparts of half a century ago, and they are also likely to own larger numbers of the same type of item. For instance, one survey found that it is now common for young and middle-aged people to own twenty or more items of products such as shirts or briefs – compared with three or four of these items in the 1940s and 1950s (Carr & Latham, 1988).

In an analysis of the attempted application of mass production methods to consumer durable products in the 1930s, the American sociologist Lewis Mumford identified a number of possible dangers. Referring to products such as cars and furniture, he took the view that 'there is great danger that once the original market is supplied, replacements will not have to be made with sufficient frequency to keep the plant running. [... Manufacturers] are driven desperately to invent new fashions in order to hasten the moment of obsolescence; beyond a certain point, technical improvements take second place and stylistic flourishes enter' (Mumford, 1945). You may well feel from your own impressions and experience that this point has perhaps been passed in the case of many clothes and other textile products. So, in the light of this possibility, what can we learn about design from the modern textile sector?

## 1.2 TEXTILES EXERCISE, PART 1

The exercise extends through to Section 4. Here, and in Section 2, you are asked to look at textile products in and around your home. You will need to apply some of the knowledge and skills you have gained from earlier parts of the course to the consideration of a number of questions. These are partly concerned with direct issues related to the design and development of the products you locate, but they are also concerned with the way that a number of 'external influences' (i.e. outside the immediate area of product design) may influence the way that design is conceived and approached. What you will get from this section, and the Block as a whole, depends on the extent of your active input at this stage. This input should be at three levels:

1. Instead of us sending you a home kit, you will have to identify your own 'kit' by determining the range of textile products in your home.
2. You will need to examine some of these products and consider the questions I ask below.
3. Most importantly: in considering your responses to these questions, you should actively think about what you have learned from your prior work on the texts and other components in Blocks 1 to 5, including the Guided Design Exercises. How can this knowledge and experience be applied to provide insights about some of the issues which designers face across the various parts of the textile sector?

**Read through the exercise, including the six questions, before starting work on this activity.**

### TEXTILES EXERCISE, PART 1

As the starting point, spend about an hour to an hour-and-a-half looking around your home to identify the various types of textile products it contains. While doing so, jot down your findings in your Workbook in the form of answers to the questions below. (If you are currently away from your usual home, then either work from memory or undertake your survey in any other context that *is* convenient.) You may find that the first part of this activity – product identification – is best accomplished by fitting it in with other, routine tasks. This will give you time to think about the progress and implications of your search.

We are concerned with a group of products here, and one of the things you will have to think about is: What products might be regarded as textiles? There are some obvious textile products such as curtains and coats, and you might find it useful to look at these initially to form some ideas on the defining characteristics of textiles. What things do 'textiles' have in common? Once you start to think in these terms, you should be able to find a surprisingly wide range of different types of product, both in and around your home. A definition is attempted in the next section if you feel stuck, but try to work one out for yourself first.

When you look around, use your design awareness to take account of the different types of textile, their surface appearance and other properties, what materials they are made of, how they appear to be constructed, and so on. Think about the issues that might have confronted the designers, whether there appears to have been any significant design input, and what elements of design might have led the original purchaser to choose that particular product. One final point: while the exercise is concerned with products identifiable in the immediate surroundings of the home, you will find the usefulness of the exercise is extended if you look more widely when you have the opportunity – at the styles, quality and so on of what other people are wearing, at other uses of textiles in the outside world, and at the way textiles are marketed in retail and other outlets.

Make notes in your Workbook as you answer the following questions:

**Question 1**
What is the range of textile products in and around your home? (As was emphasised above, you will need to think broadly about what constitutes a 'textile'.)

**Question 2**
Your survey in response to Question 1 will identify a range of clothing and other types of products. Thinking only of the items of outer clothing for the moment, in what ways do you think the product range you have identified may differ from what you might have found if you had carried out the survey (a) twenty or so years ago, and (b) forty or so years ago? What factors seem likely to have shaped clothing design?

**Question 3**
In Block 1, the notion of market 'niches' was introduced, and this was further developed in terms of market segmentation in Block 2. What is meant by these terms? What factors might contribute to segmentation in relation to clothing products?

**Question 4**
In Block 2, and in other parts of the course, concepts of 'product life' are discussed. What are these concepts? How do you think these apply to the types of textile products you might have bought during the past three or four years? What relevance do you think 'product life' might have for textile designers, and for consumers?

**Question 5**
In Block 3, and in other parts of the course, the role of innovation in product development is emphasised and, to summarise the processes which influence the direction and rates of innovation, the concepts of 'technology push' and 'market pull' are used. How significant might these factors be in relation to the types of product you have identified?

**Question 6**

Block 5 subdivides customer needs in relation to a car in the following terms:

- functional appeal
- economic appeal
- aesthetic appeal
- ergonomic appeal.

How relevant is this classification for the needs which may be associated with the main types of textile products which you have found?

**You should stop reading now, and work through the questions above before looking at the discussion which follows in Section 2. I suggest that you spend about 30 minutes on Question 1, and 15 to 20 minutes on the other questions.**

# 2 THE PRODUCTS AND THEIR CHARACTERISTICS

The questions posed in the exercise of Section 1 are broad-ranging. Obviously, in the time available, and given the diversity of products that are involved, I have had to be fairly selective in discussing the questions, and I have also had to include an element of basic information as scene setting and to convey some of the complexities of design and production in this context. So, discussion of the questions will only partly follow the numbered order. After initially establishing some basic details about the range and types of product, the rest of this section will be concerned with identifying some of the main factors and motive forces which shape patterns of product demand and their interaction with the broad task of design. One point which emerges is the inherent fragility and instability in market conditions, and the role of these in both providing opportunities for designers and imposing constraints. Some of the factors which lie beneath this instability are technological – in the form of innovations of various types – partly because innovation opens up new possibilities in production and in product designs. These factors are looked at in Section 4.

## 2.1 THE PRODUCT RANGE

### DISCUSSION OF QUESTION 1

If you have been able to undertake a fairly thorough survey in response to the first question, what may have surprised you is the magnitude of product diversity. This is apparent in terms of the types of product you are likely to have found, in the combinations of materials that are used, and in the methods of construction. Some products are of extreme simplicity. For example, a common type of cleaning cloth is formed from a single type of textile fibre (Figure 4), and it appears to consist of a single component (in fact, it is a bonded assembly of large numbers of individual fibres). On the other hand, some of your clothes, particularly the more expensive items, may be constructed from a range of materials and fabrics, and are formed from a sizeable number of components. The various products you may have identified can be grouped in three broad categories: apparel, household textiles, and industrial textiles.

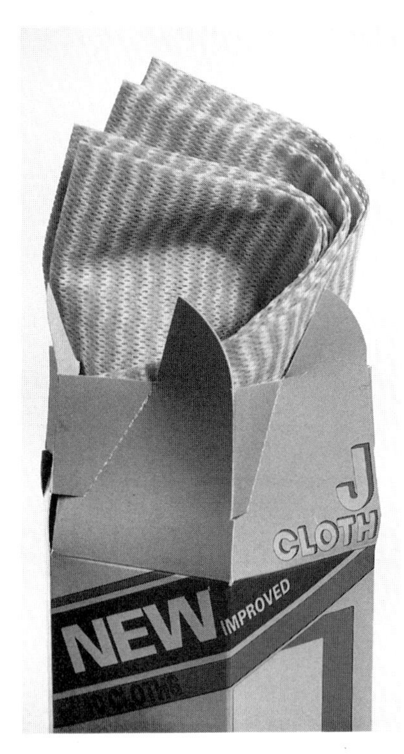


**FIGURE 4**
WIDELY USED HOUSEHOLD CLOTH
This cloth is made from bonded (non-woven) fibres – an example of a significant product innovation (for the company concerned) at a seemingly simple level.

### A Apparel

You will probably have found a wide range of products in this category – particularly if your household contains both males and females and a mix of ages. In manufacturing and in retailing, clothing tends to be divided into three obvious product groupings – women's, men's and children's wear – and there are sub-categories within these. In each category, there are particular sets of factors shaping product design. For instance, the design of women's garments is usually more fashion-oriented than men's; children's wear, though also somewhat fashion-oriented, can raise issues of safety (e.g night-clothes) and of wear resistance.

Within each of the three groupings, the range extends from fairly simple undergarments such as socks or briefs through to outer-wear. The latter may be of a considerable variety: for instance, to meet different climatic conditions (winter and summer clothing for example), and for use in various types of activities (formal, social, work, domestic, leisure, sport and so on). You may also have discovered that textile components are fairly extensively used in footwear – in items ranging from laces to the linings in Wellington boots and other footwear, the uppers of fashion shoes, tennis shoes, some 'trainers' (rarely used for training), lightweight walking boots, and so on.

(A)

(B)

**FIGURE 5**
EXAMPLES OF CLOTHING
Women's clothing (A) and men's clothing (B) found around the home.

## B Household textiles

Surprisingly, perhaps, a fairly thorough search can reveal greater diversity in this category than is found in clothing. Many household textiles fall into the following three groups:

- *Basic household textiles* – such as tea towels, pan cleaners, and a variety of functionally specific cloths for tasks such as cleaning dishes, dusting and shoe-polishing;
- *Personal items* – bed linen, duvets (including any filling of synthetic fibres), towels, blankets, disposable and non-disposable nappies, and so on;
- *Furnishing textiles* – rugs and carpets, curtains, chair coverings and wall hangings.

But you may also have found a very extensive range of less predictable items such as: adhesive plaster, bandages and other textile products in the medicine cupboard; book-bindings; dental floss; printer ribbons; caulking in stove pipes; pet's bedding; bags and cases made largely of fabric; rucksacks, sleeping bags and tents; wallpapers with a fabric surface; filters in air extraction units, ponds and aquariums; items related to gardening such as string, gardening gloves and netting used for plant protection, in conservatories and greenhouses; and ropes such as those on a child's swing.

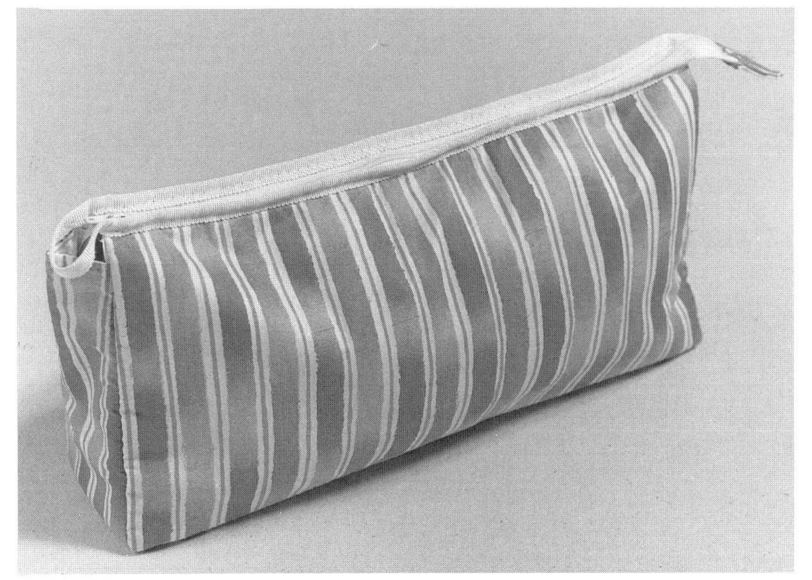

(A)

(B)

(C)

(D)

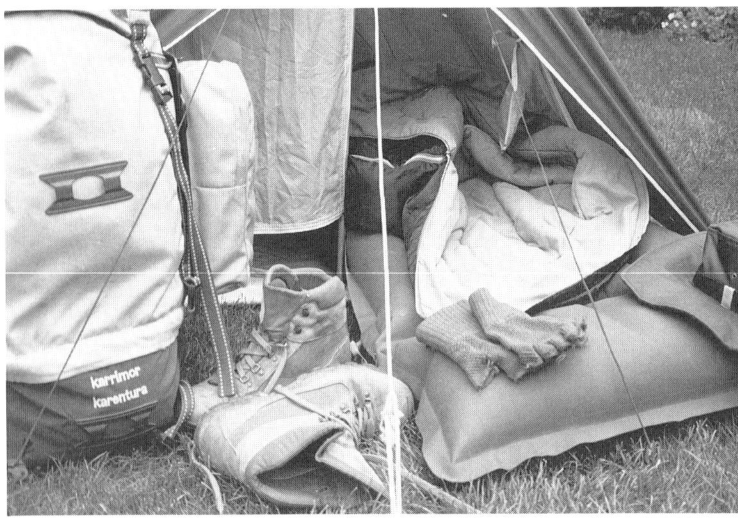


(E)

**FIGURE 6**

EXAMPLES OF HOUSEHOLD ITEMS CONTAINING TEXTILES

(A) washbag

(B) pan scrubber, and filter from cooker air filter hood (non-woven textiles)

(C) string, netting, sacks and matting

(D) less obviously, textiles are used within computer floppy disks

(E) outdoor and camping equipment also contain a variety of textiles: walking boots, rucksack panels and straps, sleeping bag outer and lining, tent panels and guy ropes, air bed panels, rain hat, gloves

## C Industrial textiles

A substantial and growing proportion of textile output (about 16% of textile fibre consumption in the EC in the 1980s) goes to the industrial sector, although only some of these subsequently enter the domestic sector, generally as components of other products. The most obvious products in this category will be found in a car: the roof linings, carpets, upholstery, other parts of the interior 'trim', and in seat belts. Less obviously, there are textile components in bicycle, car and other tyres (mostly in the radial ply of the inner body of a tyre, Figure 8). From the home kit materials samples explored in Block 5, you may also have wondered about some of the plastic composite components in racing cars (are glass and aramid fibres textiles?) and, possibly, about some other products inside your home. For instance, is the rock-wool used in roof and other insulation a 'textile'?

**FIGURE 7**
TEXTILES USED IN CAR INTERIORS: SEAT COVERS, CARPETS, SEAT BELTS, ROOF AND SIDE PANEL TRIMS, SUN VISOR COVER

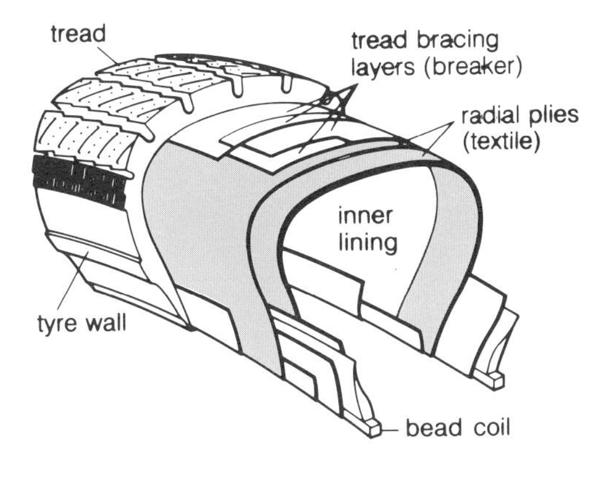


**FIGURE 8**
CROSS-SECTION OF A CAR TYRE, SHOWING LOCATION OF REINFORCING RADIAL PLIES FORMED FROM TEXTILE FIBRES

## SO, WHAT ARE 'TEXTILES'?

I emphasised above that it was important to think about what 'textiles' are. Many people probably think in terms of **fabrics**, and of products made from textile fabrics. This is certainly how textiles have tended to be defined in the past. (Note that in this Block I am confining the use of the term *materials*, sometimes used interchangeably with fabrics, solely to the technological sense of elemental substances in semi-processed form from which artefacts – such as fabrics – are formed.) But, as textile technologies have changed (see Section 4), they have enabled the development of new types of product and new end-uses – as is reflected in the following definition:

> ... originally a woven fabric but the term '**textiles**' (i.e. the plural) is now also applied to fibres, filaments and yarns, natural or man-made, and products for which they are the principal raw material.
>
> Note: this definition embraces, for example, threads, cords, ropes and braids; woven, knitted and non-woven fabrics, lace, nets and embroidery; hosiery, knitwear and other garments made up from textile yarns and fabrics; household textiles, textile furnishings and upholstery; carpets and other fibre-based floor coverings; industrial textiles, geotextiles* and medical textiles.
>
> (Tubbs & Daniels, 1991, p.312)
>
> [* Geotextiles are textile materials used increasingly in the civil engineering and similar industries for such purposes as stabilising or reinforcing, such as in road foundations or generator cooling towers (as shown in Figure 9), and in protection of the coastline, river banks, countryside, etc. from erosion.]

**FIGURE 9**
USE OF CIRCUMFERENTIAL TENDONS OF ARAMID ROPE FOR REMEDIAL WORK ON A POWER STATION COOLING TOWER

The range of products classified as textiles by this definition is probably somewhat more extensive than the range suggested by your survey. But the breadth of this definition prompts the question: What is it that these diverse products have in common?

A comprehensive answer to this would be neither simple nor short. But, as a generalisation, the underlying commonality is that textiles are a particular category of **fibres**, and the products formed from these fibres. The range of textile fibres is considered in Section 4, and includes the cotton, wool, polyester and polyamid (nylon) fibres which are likely to have been predominant in the products you examined. As you will see, there are many other types, but these have a number of common properties. These include:

1. A combination of 'fineness' with relative length. More precisely, the typical form of textile fibres is that of a very fine linear filament (with a cross-section of less than 100 μm) which has a *high aspect ratio* (the ratio of length to diameter, which is generally greater than 200);
2. *Very high tensile strength* in relation to mass (in this respect, some fibres are, for instance, stronger than steel);
3. High levels of *flexibility* (including a capacity for recovering shape);
4. Various properties of *cohesiveness* in relation to other fibres of the same or similar types.

These properties are essential for the assembly of fibres into what amount to *textile structures* of various types (see Section 4.3). In most cases, fibres are used to form *non-rigid, flexible structures*, either directly by felting or bonding, or as spun yarns which are then used in the production of woven or knitted fabrics. These basic processes are illustrated and explained in Figures 10 to 12. In some cases, textile structures are combined with resins to form rigid or semi-rigid composite materials, as in the case of the glass fibre, Kevlar and carbon-fibre weaves which you received with your home kit. However, in this Block, we are only concerned with the use of fibres in *non-rigid* structures, and primarily their use in clothing.

**FIGURE 10 ▾**
STAPLE, FILAMENT AND YARN

This figure illustrates a number of basic points about yarn formation: (1) the basic distinction between filament (very long) fibres and staple (shorter) fibres; (2) the need to align the individual fibres in parallel prior to spinning; (3) the potential for improving yarn characteristics by combining fibres of different types and properties. In this example, staple and filament fibres are combined to produce a yarn that is strong, even, and suitable for use in knitted or woven fabrics.

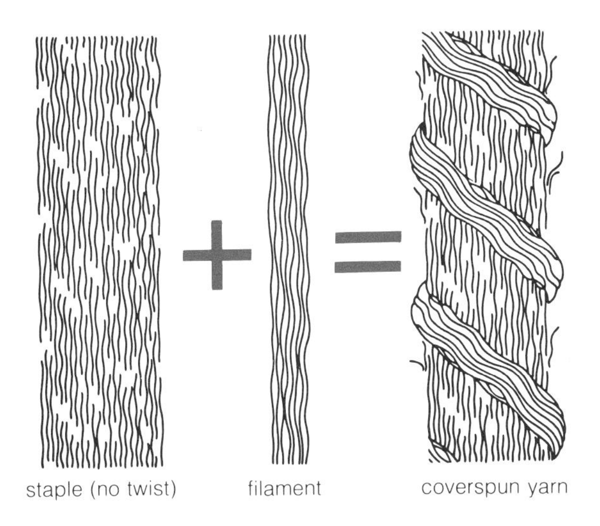

**FIGURE 11 ▸**
WEAVES

In essence, weaves are simple structures comprised of long warp yarns and short weft yarns drawn across the line of the warp. The multiplicity of fabric types derives primarily from the use of yarns of different types (in terms of fibre types, yarn weights, yarn structures and so on); the use of different types of yarn within a single category of fabric; the dynamics of the weaving process (such as loom type, the weave structure – to produce tight or loose yarns, and so on); and the wide range of combinations in warp/weft 'drafting' that are possible. Some simple examples of the latter are shown in this figure – the warp yarns are those running vertically. The application of colour by dyeing or printing adds further variations, as does an extensive diversity in finishing processes.

(A)

(B)

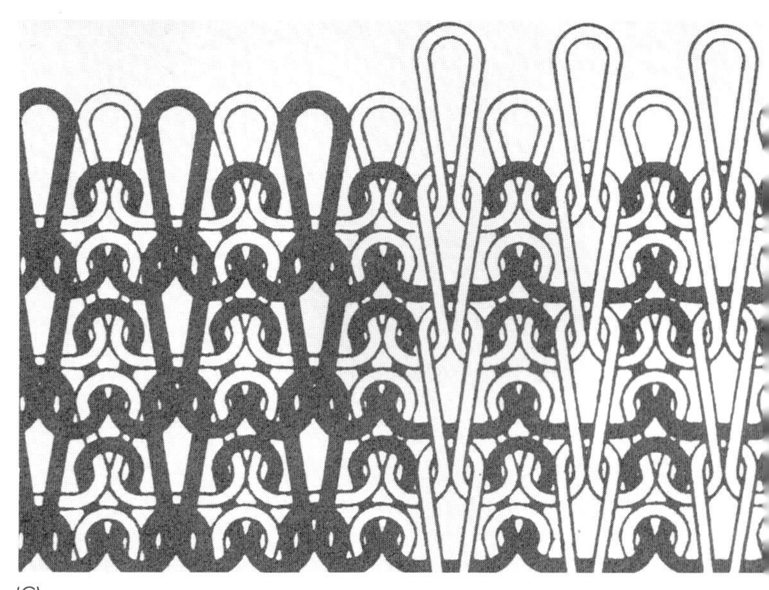

(C)

**FIGURE 12**

(A) EXAMPLES OF JACQUARD WEAVE FABRICS

(B) EXAMPLE OF DOUBLE-KNIT FABRIC

(C) DIAGRAM OF AN EXAMPLE OF A JACQUARD KNIT

(D) A JACQUARD LOOM

Manufacture of the plain weave shown in Figure 11 requires only the lifting of alternate warp threads for each weft insertion and, correspondingly, a simple loom. Complex fabric structures and patterns – whether weaves (A) or knits (B, C) demand more complex control systems. Jacquard looms such as the example in (D) have long used punched cards for control but are shifting to micro-electronic control systems and are thus ideal for direct linkage to CAD systems. (C) also illustrates the looped linkages which are the basis of knitted fabrics.

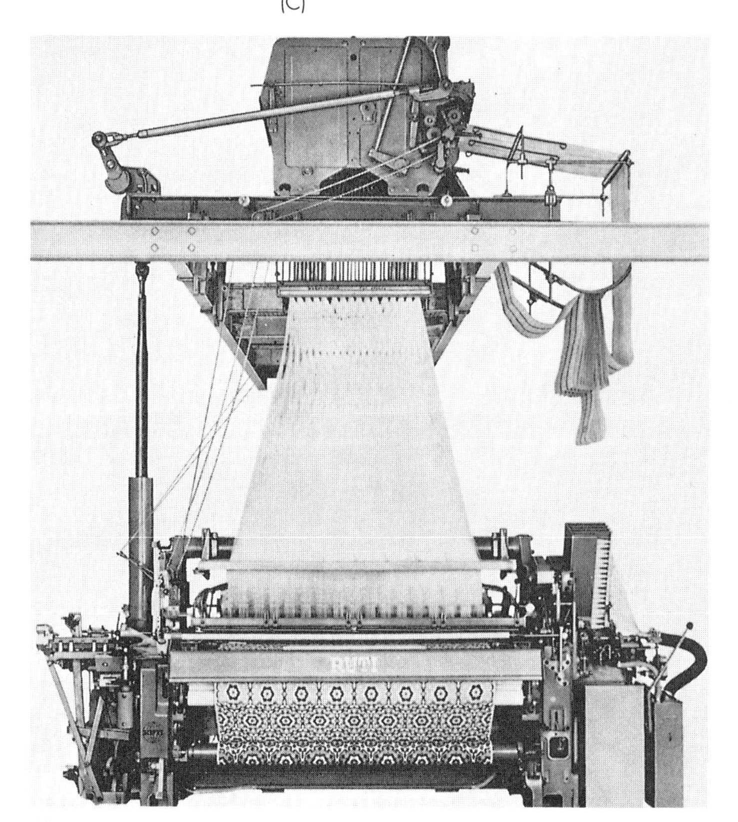


(D)

At a very different level, the technical commonalities referred to above contribute to links in terms of *innovation* and product development. Highly competitive conditions in the main market for textile fibres have contributed to pressure to find new markets for textiles outside the 'traditional' clothing and household markets, and to the development of new products for these markets (for example, geotextiles). This has sometimes taken the form of an *'invasion' of markets* for other types of materials – **substituting** textile products for those produced from other materials. For instance, textile fabrics have displaced plastics for many uses in car interiors, and have been used in new products such as emergency air bags; they are also displacing steel in uses such as the cooling tower example. In turn, such new product development has led, in some cases, to product developments in the traditional markets. For instance, 'Gortex' fabrics, now used widely in specialised outdoor wear, were originally developed for the medical textiles market. So, as in the earlier examples in the course, **technology push** and **market pull** are both central forces in driving product development – but in ways that are in some contrast to the earlier examples. We will look at technological factors in Section 4, and at market-related factors in the rest of this section.

### SAQ 1

The notion of textiles as being 'structures' in the same senses as, say, cars or buildings, may seem unfamiliar. In the light of earlier parts of the course, what does this suggest about the nature of design tasks in relation to clothing?

### DISCUSSION OF QUESTION 2

This asked about the differences which you might have found if you had undertaken your survey 20 years and 40 years earlier. Obviously, the answers cannot be precise, and the differences depend on a wide range of factors, such as people's incomes, their tastes in clothes, and so on. (You might care to think about the implications of these factors for clothing design; the point is discussed in Section 2.5.) However, a quick way of establishing what some of the general differences might be is to look through the photographs on the next four pages. Obviously, these illustrations are a personal selection from a very large range of possibilities, but I think that they convey a number of important points.

(A) 1923

(B) 1950

(C) 1950

(D) 1948

**FIGURE 13**

A BRIEF AND HIGHLY SELECTIVE VIEW OF SHIFTS IN DRESS STANDARDS AND STYLE

The two railway directors in (A) were photographed in the 1920s. (B) outside a miners' club on a 1950s Sunday morning, and (C): Edinburgh in 1951, illustrate the continuing conservatism of everyday wear at that time – although the emergence of more assertive styles of dress can be glimpsed among the Leicestershire women in (D). The parallel (to B) conservatism of middle-class dress styles during this period is evident in (E), and even among the comparatively young, in (F). By the 1970s, a far more diverse pattern of dress was evident, as is indicated by (G).

(E) 1957

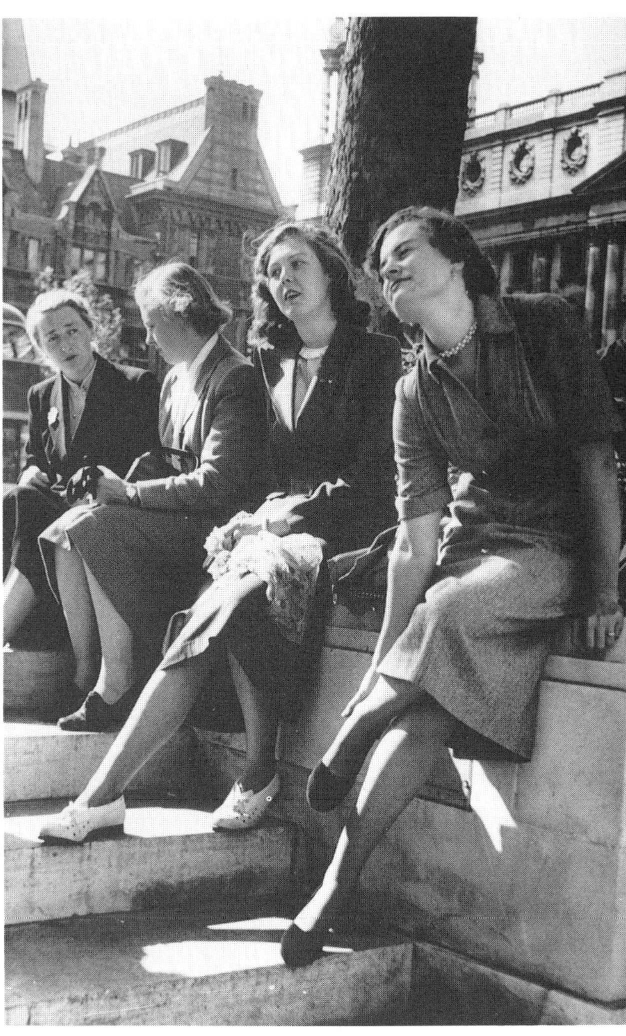

(F) 1952

(G) 1976

**FIGURE 13** (CONTINUED)
DRESS STANDARDS AND STYLES IN THE 1990S

By the 1990s, while formal dress was still required for many roles, the overall pattern was of far greater diversity in dress styles for both men and women. A much wider range of fabrics, colours, styles and fashion accessories was available. Associated with this were changes in the general quality of fabrics and clothing, and – again in general – far higher levels of laundering and other 'maintenance' of clothing. (Photographs on this and the facing page were taken in and around London, 1992.)

**SAQ 2**

Having considered the changes suggested by the photographs in Figure 13, and with your answer to Question 2 in mind, what might be the implications of these changes for clothing design?

## 2.2 DIVERSITY, COMPETITION AND DESIGN

You will be only too well aware that product heterogeneity such as you found is reflected in a correspondingly wide distribution of price levels. While basic items such as dishcloths can be obtained for only a few pence each, the more expensive items in a wardrobe – for instance, a man's suit or a woman's coat – may cost two or three hundred pounds or more, and carpets rather larger sums. This puts many textile products in a price category close to that of consumer durables such as video machines or refrigerators.

There are wide price ranges *within* particular categories. Both clothing and furnishing textiles extend into the realm of high-cost luxury goods – products such as hand-woven carpets and tapestries, bespoke suits and *haute couture* garments can cost thousands of pounds. The 'vertical' scale of price – the difference between highest and lowest priced items in a given category of products – provides one of the dimensions by which companies seek to extend **market segmentation** (bringing us to Question 3).

### DISCUSSION OF QUESTION 3

In the case of **vertical segmentation** (termed 'market segmentation' in Block 2), firms generally aim to encourage consumers to move up the product range by establishing a more finely staged gradient between products in terms of design standards, particularly quality, and sometimes by moving their entire product offering 'up-market' in the expectation that many existing customers will follow, and new ones will be attracted. In general, movement on the vertical scale is a trade-off between volume and margins – products at the lower end are manufactured in large volumes but generally yield only small profit margins for each unit produced, while lower-volume production at the upper end yields higher margins.

**FIGURE 14**
SOURCES OF PRODUCT VARIETY

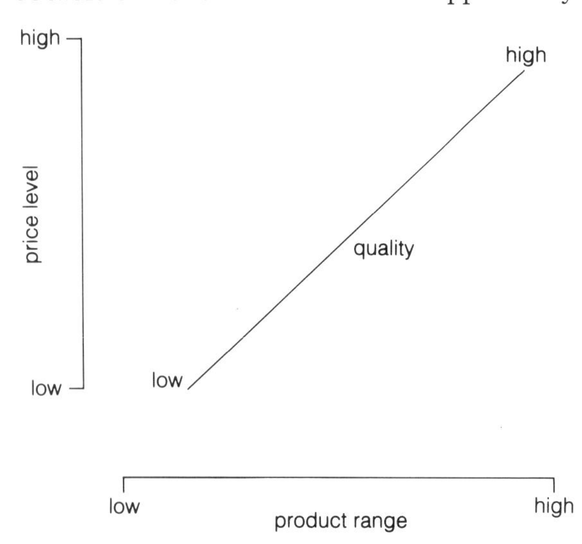


However, as your textile survey may also have indicated, firms may seek to improve their competitive position through an emphasis on **horizontal segmentation** (termed 'product differentiation' in Block 2). This involves extending the range of product choice by marketing a number of product variants within specific price/type categories – offering a wider choice of products which, in essence, are of the same kind. For instance, products of the same basic design may be offered with minor variations in cut, in different prints, and so on. The markets for most of the multitude of individual categories of products are highly segmented, and tend to be crowded by a very wide variety of product offerings – along the lines depicted schematically in Figure 14.

The extent of segmentation is, of course, an outcome of company strategies which, in turn, are shaped by market conditions. One characteristic of the latter has been the evolution of patterns of demand through the 'consumerist' boom of the second half of the twentieth century. Experience across the European Community generally shows that increased consumer affluence is initially reflected in rising shares of family income being spent on clothing, but that this share later declines. For instance, expenditure on clothing and footwear fell from about 12% of total household expenditure in the U.K. in the early 1950s to around 7% in the mid-1980s (CSO, 1980s). This is, of course, a smaller share of rising real incomes so that, over the longer term, there has still been some increase in real levels of expenditure on clothing, albeit a limited one. From the discussion of product life cycles in Block 2, you may have recognised this as indicating the stage of 'market maturity', in that demand for textile products has, in general, reached a plateau.

Slow growth in levels of product demand has contributed to an increased intensity of competition. This is because in many cases an increase in sales by one firm can only be achieved through corresponding reductions in the levels of product sales achieved by other firms. To protect their market share, firms may seek lower-cost sources of production. Hence wholesale and many retail companies have moved towards supplies from manufacturers operating in countries where labour costs are low. (By the 1980s, about 30% of the U.K. clothing market and 40% of other textiles were supplied by imported products, compared with 12% in the late 1960s. Design and some other functions are generally retained by U.K. companies which have moved their production abroad.) Since low prices are often not sufficient in themselves to hold market shares, companies also have to place emphasis on **non-price factors** in competition, notably on product quality, on developing new markets, and on extending the degree of market segmentation. In all of these dimensions of competition, design strategies and the standard of design practice are critical for success.

To a considerable extent, increasing market segmentation has followed from the strategies of retail companies, particularly those with a large share of the market. For instance, two companies accounted for about 25% of clothing sales in Britain in the early 1990s: Marks & Spencer with some 16% of the market, and the Burton Group with about 10%. But reference to retail strategies brings us to two of the more unusual features of the textile sector. One of these is that product development is only seldom a process of developing a single type of product. From couturiers to retailers, large and small, the emphasis is on the simultaneous presentation of 'ranges' or 'collections' rather than individual products. This is partly a matter of assembling a range of complementary styles for a given season. Increasingly, it has also involved co-ordination in terms of colour, 'silhouette' and so on between garments of different types, and which often extends outside the area of clothing to include a variety of fashion 'accessories'. 'Co-ordination' in this respect contributes to pressures for higher levels of uniformity in colour and other standards between the garments in production runs from the same source and from multiple-production sources. This intensifies emphasis on the rigour of design and production standards (see Section 4.3). The second unusual characteristic is the direct involvement of retailers in product design. To a significant extent, product development and design are 'retailer driven'. The role of large retailers extends well beyond the roles of distribution, display, selling and providing market feedback discussed in Block 1 (pp.102–103) to a direct involvement in product development and design which influences all areas of textile production (see below and Section 4).

> What factors might explain this 'pro-active' role of retailers? (Think about this for a few moments before reading on.)
>
> ---
>
> My answer is given in the following text.

A number of factors are important, some of which relate to the discussion of company strategies, market analysis and market positioning in Block 2 Section 6, 'Design for the market'. In the clothing sector, and more generally, these factors have been particularly evident and influential in the role of Marks & Spencer – and in emulation of M&S by other retailers, as has been described by Tse (1985). Two key strategic decisions by M&S in the 1920s were:

- to deal directly with manufacturers – eliminating the wholesalers which then dominated trading; and
- to establish their own brand name – which subsequently became, with very few exceptions, the only brand of goods sold by M&S.

These decisions reflected a fundamental philosophy of retailing in which key elements were a concentration on fast-moving (high-volume) lines, and competition based on a combination of price and quality.

Competing with the branded goods available in other shops involved M&S in the purchase of large volumes of products made to a high standard. The volume levels required were often beyond the capacity of a single manufacturer, and to ensure product uniformity, detailed product specifications had to be developed which established the cost, technical and other standards. This was well before many manufacturers had begun to operate in such terms (remember from Block 2 how critical the development of a detailed product specification is in design). It thus presented a challenge to the established (less rigorous) patterns of product design and manufacture. While there were initial difficulties, successful achievement of high sales volumes gave M&S the bargaining power to require its suppliers to accept M&S specifications as a condition of contract.

To establish and to continually improve product specifications to a standard which advanced the company's reputation required that M&S develop its own design capacities. This involved the establishment of technical departments, the concerns of which have inevitably extended from finished products to the types and standards of material used and to production technologies across the whole supply chain. In these respects, the pool of expertise held within M&S came to outstrip that of many of its suppliers – hence the often encountered description of M&S as 'a manufacturer without factories'. Inevitably, with increasing emphasis on styling in the design of clothing and other textiles from the 1950s (see below, and Section 4), so M&S's involvement and expertise in that area grew correspondingly.

The success of the 'M&S approach' has influenced the development of design capacities in other retail chains (and not only in the U.K.). Thus, the textile chain has become somewhat unusual in that many large retailers are directly involved in product development. They employ designers who are sometimes major contributors to what were termed in Block 2 the task clarification, concept and embodiment stages of design.

(However, textiles are not unique in this respect – the role of retailers, including M&S, in the technologically-led development and design of food products provides a comparable example.)

But while **retailer-led product development**, in some cases, has contributed to improved product quality, it is also the subject of criticism. Among other things, the large retailers are said to have used their market power to impose 'uniformity and compliance in design' on clothing and fabric manufacturers, to have demanded 'trivial, bland fabrics', and to have stifled the capacity of manufacturers to advance new designs (Colchester, 1991). While these criticisms may have some substance, they may underestimate the difficulties of designing and manufacturing products for a large market – points which are taken up in Section 2.5.

## 2.3 PRODUCT LIFE

### DISCUSSION OF QUESTION 4

The notion of **product life** (raised in Question 4 of the Textiles Exercise) is increasingly important for designers to understand and apply – as will become clear in Section 6. But the term has a number of meanings, all of which need to be taken into account. The term was introduced in Block 2 in the context of **product life cycles**, in which products move through successive phases of introduction, growth, maturity and decline. In addition, product life has been referred to in terms of the total life cycle – from the 'cradle' (raw material production) through to 'the grave' (its end-of-life disposal). We will consider the relevance of these in later parts of the Block. In this section, I will consider two more particular aspects of product life: the 'life' of the product as a retailed item, and its 'life' in subsequent use. In both senses, the lives of some textile product lines, particularly clothing, can be very short indeed. For instance, a survey by the U.S. Congress's Office of Technology Assessment (OTA) found that 35% of apparel products had 'in-shop lives' of 10 weeks *or less*, and 80% had 'lives' of 20 weeks *or less* – as is shown in Figure 15.

**FIGURE 15**
PRODUCT DISTRIBUTION BY RETAIL OUTLET LIFE, U.S.A., 1980S

The distribution of product lives is probably similar in the U.K. – certainly, the trend is towards the American pattern. There has been an increasing emphasis on fashion content as an instrument of competition by both retailers and manufacturers. Since the very essence of 'fashion' is that there is rapid change which both brings new products and renders others redundant this inevitably means that more products are short-lived, and that 'lives' may be getting shorter. This is indicated by increases in the number of retail 'seasons' – and thus of 'collections' – in some cases, to as many as nine per year. Additionally, retailers and manufacturers have sought to lift hitherto 'mundane' items like socks and men's undergarments from the basic category towards that of fashion as a means of stimulating demand and gaining market share.

Emphasis on short-life products has implications in terms of levels of **risk**. When any given 'season' starts at the retail end of the production chain, substantial costs have already been incurred on product development and on manufacturing a large part – if not all – of the production run. Risk is increased because of limited possibilities for test marketing, consumer product testing and market research. Experience appears to have led retailers to rely instead on five main sources of market intelligence:

- their own market feedback – which has been greatly enhanced by the rapid generation of accurate data through the use of electronic point-of-sale systems (EPOS, see Section 4);
- the forecasting of both fashion trends and future levels of demand – also aided by EPOS, and also undertaken by a number of businesses which specialise in forecasting;
- the skills and experience of their buyers or buying teams;
- the sensitivity of designers to the direction of fashion trends;
- information gained through an extensive information network which operates via an extensive trade press, trade fairs and the wider 'fashion industry'.

However, the high level of unpredictability of consumer choices and a potential for rapid changes in the overall level of demand contribute to limitations on these intelligence and forecasting systems, and this carries substantial financial penalties. Under-estimated demand results in lost revenue, such as through gaps in the availability of particular sizes in some ranges ('stock outs'). Over-estimated demand results in surplus goods which have to be offered at reduced prices in end-of-season sales. The costs of inadequate forecasting are indicated by estimates in the U.S.A. that this lost revenue is approximately equal to 15% of total turnover (OTA, 1987). Paradoxically, these difficulties increase in magnitude where retailers increase the number of seasons, as can be seen in Figure 16.

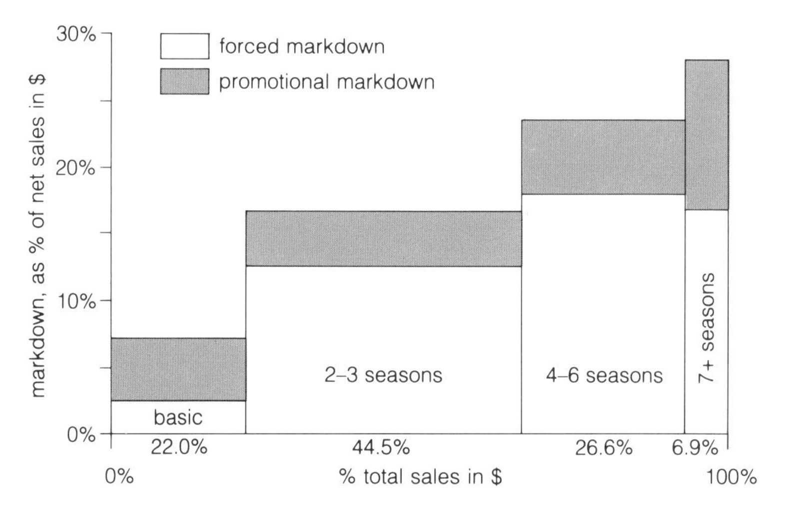


**FIGURE 16**
CLASSES OF RETAIL MARK-DOWN (% OF SALES BY SEASONAL MIX)

The design and development costs for many textile products are comparatively low, even in the larger businesses, in relation to, say, the costs of developing a new vacuum cleaner or other consumer durables. But that is not to say that these costs are negligible. First, short product life means repeated exposure to the uncertainties that generally accompany the launch of a new product. (Contrast the level of exposure for a range of fashion items with an electrical product for which a significantly new design may be launched only after a few years.) Second, development and design costs are often incurred, not for single products, but across a range of products of different types, adding to the costs of development.

The result, for the larger firms at least, is a design process which is more complex than may be supposed. Figure 17 indicates the number of stages and iterations in the *product development cycle.* A number of influences emanate from the broader fashion and production sectors, and these interact with the ideas generated by in-house fashion designers, and with

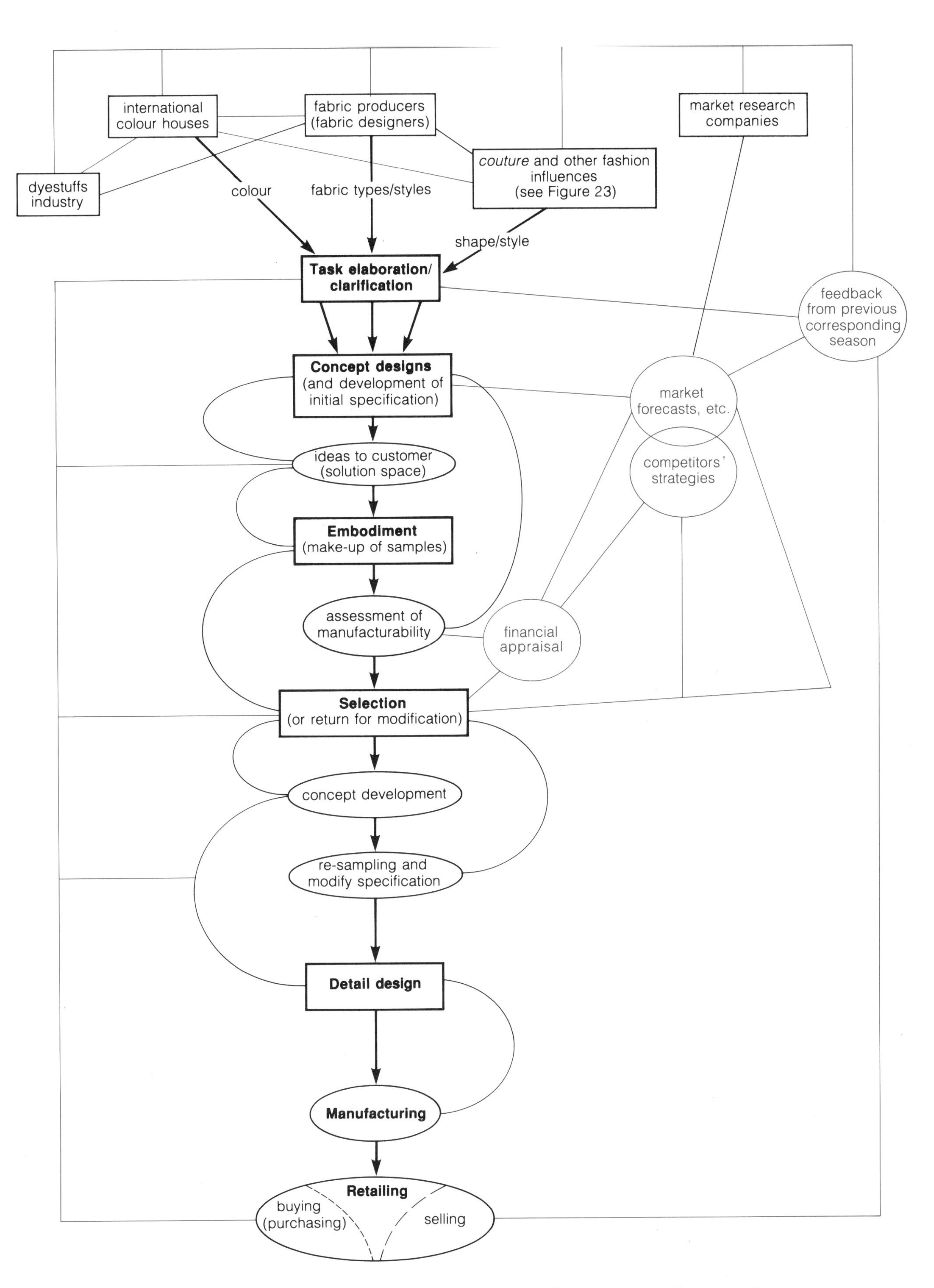


**FIGURE 17**
FASHION DESIGN CYCLE, RELATED TO THE COURSE MODEL OF DESIGN (BLOCK 2, P.118)

feedback from the corresponding season in the preceding year. Successive influences from retail decision taking, financial assessment and other factors shape the development of designs as they move towards final design commitments. The overall process of concept development, testing and modification can be compared with the model developed in Block 2.

**SAQ 3**

How would you summarise the design process depicted in Figure 17?

The complexity of the design process contributes to a total *design-to-production cycle* which is surprisingly lengthy, given the shortness of product lives and the potential for rapid changes in styles. From the initial design stage to the time of first sale, the timescale has widely been around *50 to 60 weeks* for many higher volume fashion items sold in the multiple outlets. Consequently, while you are reading this in (say) September, many manufacturers and retailers are already in the early stages of planning and designing their product ranges for the winter *after* next! Even small independent retailers who buy their collections from wholesalers face considerable lead times between placing orders and their delivery. The implications of this, even for supposedly standard products such as jeans (where style changes are generally limited and the development cycle is correspondingly short) are

> ... crazy. It forces retail buyers to guess four or five months down the road what a consumer who is 15 years old is going to want in jeans. During that time, a new movie can come out, and the trend goes from blue denim to black denim. And suddenly the inventory commitment is obsolete, causing costly mark-downs.
>
> (Howard, 1990, p.136)

**SAQ 4**

From what has been said so far, what factors might contribute to a lengthy development cycle in clothing products?

The length of the development cycle stands in very stark contrast to the shelf life of many clothing products. In contrast to rapid fashion changes, there is a rather ponderous process of development and manufacture. As you will see in Section 4, a major concern is to drastically reduce the cycle time through changes in methods of manufacture and through changes in design technologies and methods.

**SAQ 5**

On the basis of the discussion so far, would you classify the pressure for the development of new clothing products as arising from 'market pull'?

Finally, a brief reference to a further meaning attached to 'product life' – that of the post-sale 'life' of an individual product. In this respect also, some textile products have very short lives, for instance, some of the medical textiles used in the home and items used in household cleaning. For some people, clothes bought for their fashionableness may also be 'short life' – worn for a brief period before they begin to seem 'dated' and are consigned to the dustbin (although they may be recycled as decorating or gardening clothes, or via a jumble sale or charity shop). But most products are purchased – and manufactured – with rather longer lives in mind. A suit or an expensive jacket may be expected to last for several (or many) years. Equally, many furnishing textiles have potential lives which are considerably longer than those of many domestic mechanical and electrical products.

Two points follow from this. First, the *semi-durable* nature of many textile products permits the postponement of replacement buying, a factor which can exacerbate the impact of economic recession for manufacturers and retailers, and which adds to the uncertainties of predicting levels and patterns of consumer demand. Second, most products have to be designed to a standard which is acceptably hard-wearing, permitting use over long periods and which, in such respects as shape retention and colour fastness, is resistant to the rigours of use and to repeated machine washing or other cleaning. As you might expect, such factors are important in the technical aspects of specification.

## 2.4 MARKET ACCESS

Sections 2.2 and 2.3 have identified some key characteristics of the market and competitive environments in which textile designers have to operate. If you think back to Block 5, some similar characteristics can be found in the market for automobiles. Consumer demand in the industrialised countries is close to saturation levels. Intense competition between manufacturers has led to a similar emphasis on product innovation, and to strategies aimed at the extension of market segmentation. In all of these, the stylistic and technical dimensions of design are fundamental. Designers and design teams need to maintain contact with a wide range of rapidly evolving possibilities, such as in materials technologies. But some American and European car producers were slow in responding to changing competitive conditions, and were subsequently not very successful in adapting. These difficulties can derive, in part, from the considerable obstacles to design innovation that are often encountered, particularly in large organisations, from a lack of openness to new ideas and possibilities. For instance, established firms usually have an accumulation of financial and personal commitments, particularly to past capital investments, product technologies and manufacturing methods (for instance, remember the example of the 30-year life of the Austin 'A' series engine in Block 5). In general, the newer car manufacturers in Japan and elsewhere were able to develop competitive and design strategies unencumbered by the same weight of past commitments and attitudes.

So, established firms may tend to be inflexible, but it is often difficult for new enterprises with fresh ideas to enter an established market. There are considerable barriers, such as those of capital requirements and the competitive strength of existing firms. (In the case of the newer car firms, assistance from financially strong parent companies and from governments has been a factor.) But the potential for new firms to enter a market is important, for reasons which you may recall from the discussion of business strategy in Block 2, Section 6.1.

**SAQ 6**

In what ways was it suggested in Block 2 that potential new entrants played an important role?

In some parts of the textile sector, for instance some areas of weaving and knitwear production, garment making and retailing, there is a high potential for the entry of new firms. The barriers against **new entrants** are low, partly because the capital requirements can be very limited, and because work can be undertaken in small units, often outside the formal economic and employment systems. For instance, in skilled hands, garments can be made with little equipment other than a sewing machine, patterns and fabrics. Basic garments such as tee-shirts can be purchased as finished basic items and converted into higher value 'fashion' items by using simple screen-printing equipment. The results of these endeavours can then be marketed in the street, on a market stall, or in a small boutique. The significance of such 'cottage industry' products should not be underestimated, despite the predominance of the large retail chains. Not least, new entrants with bright ideas potentially provide the foundation of much larger concerns. As an example of this, Fiorenza Belussi describes the origins of the highly successful Italian-based garment manufacturing and retail franchising company, Benetton, which is now a large multinational enterprise trading under the name United Colors of Benetton:

> Benetton's story began with Luciano Benetton and his sister Giuliana. A difficult domestic situation obliged both to work while still very young: Luciano was a shop assistant in a textile shop in Treviso and his sister was working in an artisanal knitwear producing factory. Luciano was also moonlighting as a salesman of a knitwear firm from Carpi (Carpi is a specialised ancient area of knitwear producers). In the 1950s the production in this sector was highly decentralised (and focused on the role of the commissioning buyer who, on the basis of a sample collection made by the company, collected orders for shops and subcontracted work to the home workers). In 1957, they decided to work together. Giuliana had discovered a talent for designing and making knitwear; Luciano would collect orders and Giuliana would produce them at home. So, the origin of Benetton's organisational structure lies with the ancient local putting-out system, which was never fully superseded by the factory mode of production. In 1965, they established a small factory in Posano (in the Venento region of North-East Italy) with 60 employees. At that time, the other two brothers (Gilberto and Carlo) joined the company. The division of labour among the family was clear cut and it is the same nowadays: Luciano deals with the marketing, Giuliana with the design function, Gilberto with administration and finance and Carlo is in charge of production.
>
> (Belussi, 1987, pp.10–11)

Similar stories are encountered in Britain. The best known of these is probably that of Michael Marks, whose experience as a street peddler – of haberdashery – and then as a market stall holder led to the foundation of Marks & Spencer. Mary Quant, Zandra Rhodes and Laura Ashley provide similar well known examples (the latter started by screen printing designs for table mats and scarves on the family's kitchen table). There are many others whose businesses, although successful, have thus far remained small-scale, plus countless others who try, and fail. But it is not only small firms who make up the new entrants to the market. Over a number of years, a number of large supermarket chains whose main focus was on food retailing have expanded their range into clothing. Clothing and other textile products have also proved to be useful targets for mail-order firms.

First, as you may have expected from the discussion in Block 2, accessibility for new entrants is a *major* factor in the openness of parts of the textile sector to innovation in product types, and to the dynamism of the sector reflected in new print designs, garment fashions, and so on, and in the emergence of new market niches (see the example in Section 2.5). Second, low costs of entry have contributed to a tendency for the supply of textile products to exceed demand, contributing to the uncertainties and high risk levels of the overall market for clothing (and other textiles). But this is an international, and not a domestic phenomenon, not least because a substantial part of the market is supplied from Third World sources – often through the agency of U.K.-based firms which, while retaining design and other capacities, sub-contract to overseas producers. This has contributed to downward pressure on production costs and thus on prices, so that the prices of many textile products have generally fallen continuously in real (constant price) terms. The cost-minimising role of design has been correspondingly emphasised. But downward price pressure has also contributed to an increasing emphasis on **non-price factors** in competition, such as quality, design and after-sales service. (The role in consumer choice of factors other than price is referred to in Block 2.) These conditions are a further factor which contributes to emphasis on design innovation and change. As you have seen, *fashion* is an important element in this, and it is considered next.

## 2.5 FASHION AND STYLE

Some of the products found in your survey will doubtless be of a rather humdrum character, perhaps because of their association with the endless process of maintaining some semblance of order in the domestic scene, or because of their lack of aesthetic appeal. Their functions are, in the terms of Block 1, essentially *practical* or *utilitarian* – or so it may seem. But, as Figure 18 illustrates, in the drive to develop distinctive products and thereby influence consumer choices, even products like the humble oven glove can be elevated towards the category of fashion items. If you then consider the wider range of household textiles in your survey, from sheets to curtains and upholstery, the visual and related elements of surface design – colour, patterning, shape, tactility and so on – may appear to be predominant, as is illustrated in Figure 19. In the case of clothes, these attributes can be far more important. Here, we enter the realm of what Block 1 termed **psychological functions**.

**FIGURE 18**
CONTRAST BETWEEN SIMPLE FUNCTIONALITY (A); RUGGED FUNCTIONALITY (B); AND THE NOVELTY DRIVEN EPHEMERA OF (C) AND (D)

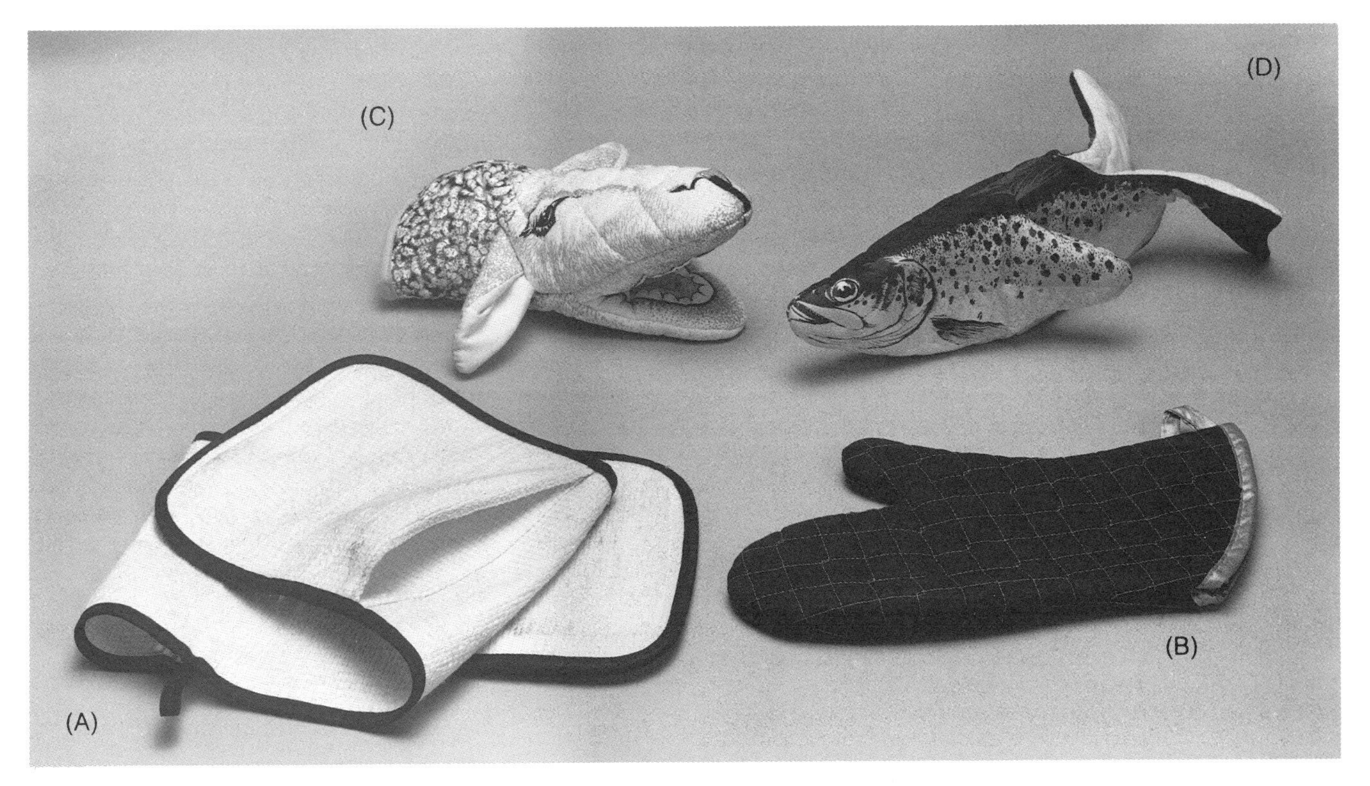

**FIGURE 19**
EMPHASIS ON PATTERN AND TACTILITY IN FABRICS

**SAQ 7**
What were the 'psychological' motivations underlying the design of a product identified in Block 1?

These functions are all potentially important for textile designers, as is particularly evident when the general ambience of fashion and style is taken into account. But their significance as motivations on the part of designers varies greatly. Consider, for example, the differences between *haute couture* and less exotic, '*mainstream*' areas of the market. Reflecting the basic division of craft and mass production in Block 1, mainstream designers have to attempt to reconcile the varying demands of large numbers of people while *haute couture* designers generally produce for individual customers, and collectively cater for a very small total clientele – possibly as few as 3000 in the world (Coleridge, 1988). Clearly, as is illustrated by Figure 20, design at this level is very much a matter of mood shaping and of statement. An indication of this is given by the following account of the work of François Lesage whose embroideries:

... were particularly admired by the thousand or so women in the world that could afford them, since they represented the pinnacle of exclusivity – hand-embellished fabrics for hand-made clothes. Lesage remains the leading supplier of embellishment to the Parisian couturiers. Each season, he provides them with a 'palette' of materials, patterns and embroidery styles which he hopes will anticipate the mood of their collections. After lengthy discussion with Lesage the couturiers present their sketches which Lesage and his team of embellishers must then translate. Lesage's skill resides in his capacity to provide innovation each season within a restricted and conservative idiom. He must embellish fabrics using constantly new and enticing combinations of materials: raffia, semi-precious stones and leather one season, cockerel feathers, beetle wings and Czechoslovakian jet the next. [...] By 1987 Lesage's embroideries had come to be seen as the hallmark of exclusivity, a metaphor for [the era of] conspicuous consumption.

(Colchester, 1991, p.14)

**FIGURE 20**
FASHION DESIGNER AND DEMONSTRATOR – 'MOOD SHAPING' AND 'STATEMENT'?

Despite its supposed exclusivity, *haute couture* is linked to the mass market in a number of ways. In part this is through the multi-million pound businesses that have been built around the marketing of less exclusive products trading on the design name. But *couture* also influences the styles that emerge in 'mainstream' design, and in other areas. Yet, while *couture* is often projected as a fashion leader, the reality is more complex. All sectors are influenced by what is happening in their counterparts, and styles tend to emerge from a dynamic, interactive process involving a variety of sources. This process has been greatly speeded up by the widespread use of CAD in fashion design, which means that designs from any one area can be rapidly copied and re-interpreted for another market segment.

The dynamics of fashion change derive from a number of sources. These include a number of specialist design houses whose ranges are produced in small numbers compared with the mass market, but whose ideas and interpretations of trends influence the mainstream. In addition, there remain a significant number of manufacturers with established brand names whose designs, often in specialised areas, also contribute to the range of styles that are available at any particular time. Less obviously perhaps, the design of sports clothes can influence trends in styles and in underlying designs. Developments in fabrics and garment types which have been undertaken specifically for the sports market have sometimes been recognised by designers in sectors from *couture* through to mainstream fashion as offering new possibilities. The most obvious example here is the use of stretch fabrics such as Lycra, first in sports apparel and then in a variety of fashion clothes.

But sports products have sometimes been 'hijacked' by fashion-conscious consumers, emphasising that the movement of ideas, statements and styles is not solely 'top down'. More generally, there are strong, sometimes very strong, 'bottom up' influences from street culture. From 'Teds' through to punks – see Figures 21 (A) and (B) – a wide range of street cultures have been made manifest in very different fashions and in the wider elements of sub-cultural styles. These fashions are sometimes expressions of mood and social statement that manifest a challenge to prevailing values and attitudes and are in sharp contrast to those of *couture* and other 'establishment' fashions – see Figure 22. So the styles at any one time emerge from a complex set of interactions between different market areas, from the wider 'fashion industry', and from the flow of new ideas originating in the education sector from fashion students and new graduates. Some idea of this is given by the influence of street styles in the 1980s when a number of fashion design graduates and students set up businesses which were oriented to street-style fashion:

> The more subtle designs exhibited a tendency towards self-parody that tapped into the nation's sense of humour and provided a good record of the 'no future' generation's first contact with the commercial world. [....] Both Helen Lipman, working for the fashion company English Eccentrics, and the young print-design partnership Hodge and Sellers, made mock of the nascent craze for designer clothing with print designs made from famous people's signatures and the reverse side of couturiers' labels – a wry parody of the growing trend of sporting a designer's or couturier's label as a means of accruing status. [...] The confusion of contradictory styles and looks that resulted from this influx of independent talent in textile design was artfully marketed by the menswear designer Paul Smith, who made a medley of strong patterns into a fashion look.
>
> (Colchester, 1991, p.16)

(A)

(B)

**FIGURE 21**
PIONEERING 'STREET STYLES'

(A) To fully appreciate the significance of style here, look back to Figure 13(B). 'Teds' were radical pioneers in the emergence of a new assertion and autonomy in affluence, style and attitudes – primarily among young working-class males. Their influence continued, as in the case of these 1970s followers.
(B) Punk similarly manifested a sundering from the prevailing conformity.

(A)

(B)

**FIGURE 22**
FASHION INDUSTRY CONCEPTS

(A) 'Mad Max look' from the 1991 Spring/Summer ready-to-wear collection by Kansai Kamamoto. The garment is a vinyl blouson adorned with metal sheets and Plexiglas darts (not all clothes are made of textiles).
(B) Male styles have also increased in significance on the fashion catwalks – even if less so in popular style. This outfit is from Matsuda's 1990 Paris Spring/Summer men's collection.

Concentrating on one area, what types of factors are likely to shape 'market acceptability' in mainstream clothing design? (Think about this for a few minutes.)

---

Obviously, the answers to this vary between designers and between the organisations they work for as employees or consultants. Some of the main factors are identified in Figure 23. One set of factors are organisationally oriented, and you may have anticipated them from earlier parts of the course. They derive from such diverse concerns as the firm's current strategy of market positioning and the associated parameters of cost, 'house style' and quality. Equally important is the way that past experience – as reflected in sales patterns – is interpreted by buyers, buying teams and others involved in the processes of decision-making around the design area.

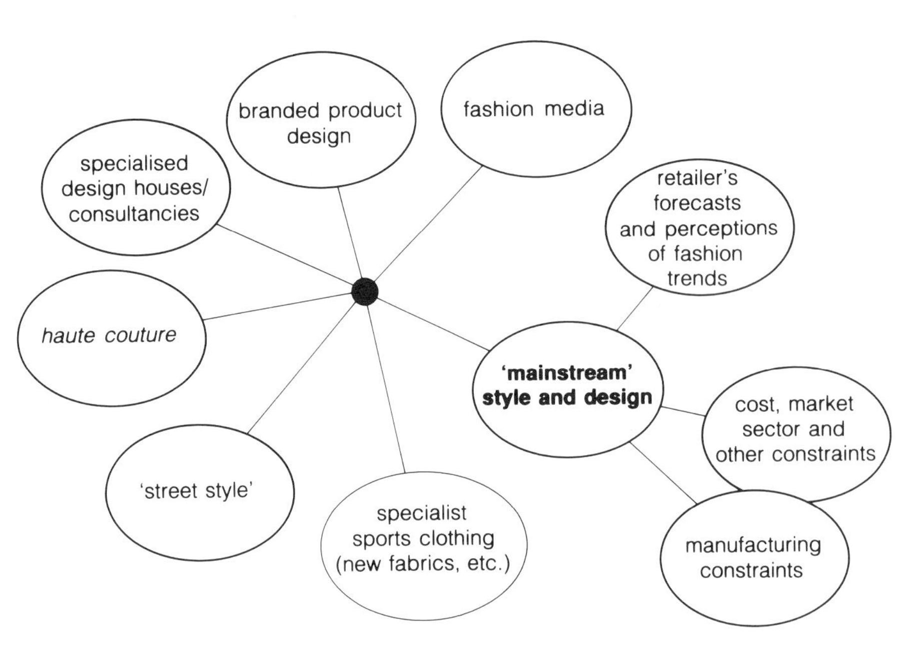


**FIGURE 23**
SCHEMA OF MAIN INFLUENCES ON MAINSTREAM FASHION DESIGN

The second set of influences – which are particularly evident in the design of women's clothing – are those deriving from the interactions between the various groups in the 'fashion milieu' considered above. These are shown grouped around a 'hub' in Figure 23. In the interactive process of generating new ideas and styles, mainstream designers are more likely to be interpreters than initiators of fashion. At the most hostile level, designers in large manufacturing and retail concerns (like those in the myriad small firms operating in the back streets of large cities), are often dismissed as mere producers of 'knock-offs' – re-interpreters, if not copiers, of the original, innovative products of designers working in other sectors. There is some accuracy in this view. Even if there is not a direct lift of garment styles, the round of fashion events has a strong influence. Buyers from large retail concerns, manufacturers, and producers of trade magazines aimed at small manufacturing concerns form part of the 'congregation' at the major fashion shows held in Paris, Rome and other capital cities. Journalists from the fashion media – magazines and newspapers – attend the same events, and their view of these events also contributes to the collective view of probable trends in terms of garment shapes, silhouettes and colours, and their emergence in the mass fashions of a year or so later.

But to view this as solely a process of 'knock-off' is surely mistaken. In part, such a view reflects a preoccupation with surface style alone. Most garment manufacturers and retailers have to produce designs that are a synthesis of style or fashion trends with other concerns. This synthesis includes **manufacturability**, which imposes a number of constraints on design where volume production is concerned. You will be aware of these in general terms from earlier parts of the course, and their specific character in textiles is considered further in Section 4.

**FIGURE 24**
ASCOT, 1981

A further factor in mainstream design is customer acceptability. The world of *couture*, and its reflection in the fashion columns, is far removed from the realities which many consumers face. This is only partly a matter of economics – of what can be afforded. For many people, the fashion industry peddles images that are based on atypical body shapes through the use of models who are mostly drawn from a very narrow age band (although a few firms are now moving away from this imagery in their own catalogues). The imagery projected on the catwalks of the fashion houses and in advertising is of lifestyles which, if they exist beyond the realms of televisual and other fantasy, are far removed from the experience of most people. The task of interpretation to produce something that, while stylish, is also acceptable and 'sensible' (the term can have positive as well as pejorative connotations) is thus a highly sensitive task.

But, to approach this task, it is important to have some feel for the varied functions that clothes and other textiles fulfil. We consider this in Sections 2.6 and 2.7.

## 2.6 FUNCTION AND SYMBOLISM

**FIGURE 25**
DECORATION AND DISPLAY

Rudofsky (1947) saw clothing as having three functions – the protection of modesty; protection against the climate; display. Here the first two functions are disregarded, even though the Patagonian tribesman in the illustration lives in a severe climate. As the illustration also demonstrates, clothes are not strictly necessary for display.

The non-utilitarian functions of both clothes and home furnishings are highly complex and, of course, they vary greatly both in type and in their significance between individuals, between groups, and between cultures. To the extent that economic circumstances and individual preferences permit, one of their functions is that of providing a means by which individuals may seek to project, to exaggerate or to disguise aspects of their personality or individuality, both in the way they furnish their home and in the clothes they wear for different occasions and circumstances. They provide means through which people variously contrive to transmit **signals** – honest or misleading – about such varied attributes as motivation, mood, values, taste, power, status and wealth. This is strikingly demonstrated in Figure 24, while Figure 25 demonstrates this through an emphasis on decoration at the expense of the protective and other utilitarian functions of clothing.

There are links here with Block 4, where you considered the various possibilities for the design of semi-detached houses. One of the questions this raised was why houses constructed by speculative builders often do not accord with the '*ideal plan*', thus incurring a number of disadvantages. One of the suggested answers, you may recall, was that questions of status and image carried more weight than functional practicalities. In understanding the implications of such considerations for design, it is important to recognise that individual motivations are located within, shaped by, and directed at, wider social groups. A variety of group-related motivations shape people's selections of artefacts, both singly, and as assemblies of possessions. In the case of clothing, like other assemblies of artefacts, they 'present a set of meanings, more or less coherent, more or less intentional. They are read by those who know the code and scan them for information' (Douglas & Isherwood, 1980). An indication of the significance of 'knowing the codes', and of the extent to which, for the majority of the populations of industrialised countries, this constitutes a disjuncture from the poorer, more stable societies of the past, is given by the following account of a woman migrant from Russia to the U.S.A.:

> In Russia she wore the customary clothes of the poor, made crudely out of scraps of cloth. 'It was not a question of style' she explained, 'but of how to cover one's body in those days'. When she came to America in the 1920s, one of the first items of clothing she bought was an item she had admired from afar while still in Russia: a leather jacket, 'a symbol of both the Revolution and elegance'. Yet when she wore this prized acquisition while looking for work, she discovered that employers would not hire her; they, too, understood the political connotations of the jacket. They were afraid its wearer might cause problems, as a unionist or an agitator. Employers were looking for signs of docility, adaptability and conformity in their prospective employees. Finally, the woman was forced to make a concession to the codes of the job market: 'I dressed myself in the latest fashion with lipstick in addition, although it was so hard to get used to at first that I blushed, felt foolish and thought myself vulgar. But I got the job'.
>
> (Ewen, 1988, p.77)

But the signals conveyed by apparel have greatly varying significance. Many people have a number of roles – as employees or employers, as the temporarily leisured, as sports supporters, and so on. In some circumstances, wearing a particular set of apparel may result from formal or other requirements related to paid work or other duties. The prescribed style of clothing, whether as a formally designed uniform or as an 'expected' style and standard of dress (also a 'uniform') follows from organisational concerns with maintaining hierarchy and discipline, and with denoting an association – if not commitment – to defined organisational functions and objectives. But this type of association is often qualified. For instance, occupational groups may exert a degree of autonomy and, in part, this may be signified – and asserted – by variations in dress which constitute an occupational style.

Outside the circumstances of paid and other work, clothing is widely used to signal membership of particular groups, thus suggesting commitment to a specific set of values and other group attributes. Correspondingly, as you saw in Section 2.5, apparel may also indicate – or be interpreted as indicating – non-membership or rejection of particular groups and values – a type of stylistic 'body armour', most strikingly displayed in the apparel of the young urban unemployed. More generally, the range of roles that people may fill, and the diversity of groups they may belong to, has tended to increase. Attention to this proliferation of roles has offered attentive designers the opportunity to develop new niches through the development of increasingly specialised types of clothing.

The points made above provide only a very brief reference to the highly complex issues which underlie many of the forces shaping both design and patterns of demand. It needs to be emphasised that the social and psychological dimensions outlined above are, of course, by no means confined to apparel and other textile products, nor to their related accessories (shoes, jewellery, ornaments, etc.). They can be seen, for example, in the marketing of products as diverse as cigarettes and cars. For our concerns, one of the most important characteristics of both styles and symbolisms is that they are frequently vulnerable to, if not besieged by, broader social and economic changes, and may thus be subject to repeated modification, replacement and abandonment. This follows from a process that is at least partly conscious, whether on the part of the defenders of the symbolistic status quo, or through the medium of fashion change.

Manufacturers of textile products have long recognised the dynamics of this process as a means for stimulating or reviving demand. For instance, in the mid-nineteenth century:

> For the middle-class market, the cotton printers produced patterns, mostly on the more expensive types of fabric, that were calculated to attract well-to-do customers by the refinement and quality of the designs, as well as by their novelty. A constant succession of new designs was produced in small quantities for middle class women who wished to be dressed in patterns that they could be sure had not yet been reproduced on the cheaper fabrics worn by working-class women. [...] many fashionable designs were subsequently reproduced by the manufacturers on cheap cotton, a practice which both attracted working class customers wanting to follow the fashion, and caused the owners of dresses in the first, expensive printing of a pattern to discard them, because they had become 'common', and to buy new ones.
>
> (Forty, 1986, pp.74–75)

During the last fifty years or so, the opportunities for generating changes in styles and for their percolation across the fashion-buying population have proliferated for a variety of reasons. The main factors have included, until recently:

- a growing proportion of the population, particularly those under 25, who command the resources for non-basic personal expenditure;
- shifts in social relationships between classes and groups;
- changes in the nature of people's work and other roles; and
- people's ability, in terms of time and resources, to fulfil a larger number of roles.

These changes have assisted a momentous expansion in the range of products which, to varying extents, have become fashion-oriented, and in the range of products available in most categories. Even clothing for adult British males, a stronghold of sartorial conservatism, has succumbed to some extent, not least in the area of prescribed workwear where designer 'names' have been used for the projection of corporate image via blue- and white-collar uniforms. The significance of these changes in terms of marketing and design is perhaps best illustrated by the stratagem of some retailers during the 1980s. They sought to increase turnover by selling what were heralded as 'lifestyles' – complete 'off-the-peg' assemblages of clothes and accessories – which, presumably, appealed most to the socially uncertain or insecure. In general, the combination of fashion changes, role proliferation and people's desire for individuality contribute to an almost infinite potential for market segmentation.

It needs to be emphasised that fashion and the symbolic dimension of textile and other artefacts are, to some extent, distinct phenomena; and that there are limits to the manipulability of both. As was emphasised earlier, the flow of new ideas and styles is not solely a 'top-down' process, driven by the fashion industry. The selection of and attachment to particular symbolisms by social, occupational and other groups, and their reflection in, say, certain types, qualities or styles of clothing, generally includes a robust element of inertia, indifference to, or reaction against, fashion 'hype'.

Fashion and style may be seen as reflecting a concern with surface form. Obviously, this is correct, but only in part. The increasing fashion content of textile products as a whole, particularly in the volumes and at the price levels that have been necessary for mass markets, has depended upon developments in production technology and in the materials from which textiles are constructed. For instance, the development and improvement of elastomeric fibres has allowed designers to develop radically new types of contour-hugging garments and of leisure wear. Changes like these have mostly been long-term, following from some **radical innovations**, but also from **incremental, cumulative developments** along the pattern discussed in Block 3 in respect of bicycle design. As in the automobile industry (Block 5), design changes at the styling level are underlain by more fundamental changes in the **core technologies**. The latter have also contributed to improvements in product performance at the level of the practical functions of textiles.

## 2.7 'PRACTICAL FUNCTIONS'

Many of the more basic products found in your survey, such as tea-towels and other items used in routine household tasks, may manifest minimal design content. The predominant concern in product development and in production organisation will almost always be to minimise production costs. But basic products are generally highly functional, and consumer choices are also shaped by the standard of performance of these functions. Thus, even basic domestic items potentially incorporate a number of specifiable **technical qualities**. For instance, performance follows from properties of textile fibres in such respects as strength, lint retention, heat resistance, water absorbency and non-reactivity with the variety of chemical compounds used in household cleaning. Even seemingly subjective characteristics such as how products feel to the touch when used are underpinned by definable **properties** of materials (see Block 5). Within limits, these are specifiable through the selection of different types of fibre and fibre processing, together with specific modes of fibre assembly in a yarn, and in a weave or knit. Hence, even long-established basic products remain open to improvement through technologically-based design innovation (for example, see Figure 4).

In the case of higher-grade textiles used for furnishing and similar purposes, various technical properties similarly underpin the dimensions of surface appearance, for instance, the range or brilliance of fabric colours, the thickness of a material, the hang or drape of a particular type of fabric, the patterning or other surface characteristics which derive from particular yarn types and weaves, the surface 'sheen' or reflectivity of a fabric, and its 'handle' – how it feels and behaves when handled. Other sets of properties contribute to the performance of a fabric in use. An excellent aesthetic design is of little use if the fabric fades in sunlight or the colours run in the wash. Other factors such as shape retention, the ability to withstand heavy use, general robustness in washing or dry-cleaning, stain resistance and fire retardance, are similarly important.

When you considered the functional elements of clothing, you may have thought about characteristics such as comfort and freedom of movement, plus those related to climatic variations – warmth, coolness and protection from wind and rain. Looking more widely, a number of other practical considerations become apparent. For example, most types of clothes need to be able to withstand considerable stress, wear and tear, while retaining a reasonable appearance. This requires that garments are *engineered* in the design process, in the full sense of the term – as in Figure 26. This is especially important where clothing has more vital protective functions. For instance, in the health service, medical clothing contributes to a safe medical environment and helps to provide protection against infection. In other occupations, clothing has to protect the wearer against hazards associated with heat, chemicals and other dangers. In such cases, a tight technical specification is required. The specified standards of performance become more extensive and rigorous in the case of some occupations within the fire-fighting, police and military services: to meet requirements such as resistance to heat, toxic gases, acids, explosions and high acceleration forces.

In the area of consumer products, technical specifications are similarly becoming increasingly demanding for the specialised garments associated with a variety of outdoor and sporting activities – from swimming and cycling to skiing, fell-walking, mountaineering and caving. Perceived standards of technical performance related to the demands of improved competitive performance or safety and protection have become increasingly important in market development. For instance, in competitive cycling and swimming, product designers have sought an optimal combination of aerodynamic or hydrodynamic efficiencies with maximum freedom of movement (see Figures 27–28).

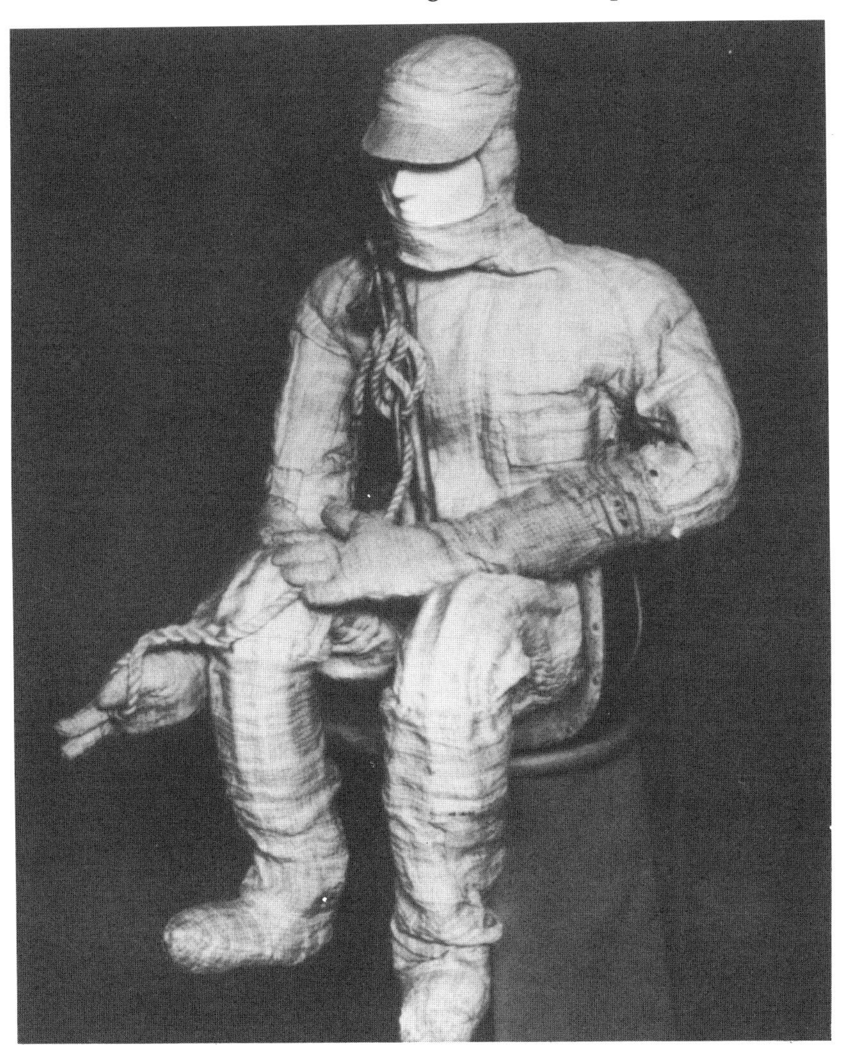

**FIGURE 26**
PROTECTIVE SUIT FOR HIGH-VOLTAGE WORK

This suit for working on live high-voltage lines was developed in 1980 and uses a specialised woollen fabric which incorporates stainless steel fibres (wool 75%; stainless steel, 25%). Obviously, its properties are defined by considerations of safety – including the greater flexibility provided by the fabric compared with earlier generations of fabric, which used cotton and steel mesh.

**FIGURE 27**
SPORTS/FASHION LINKAGES

Fabrics incorporating highly flexible elastane fibres (see Section 4) have contributed to some dramatic developments in apparel for cycle racing and other sports.

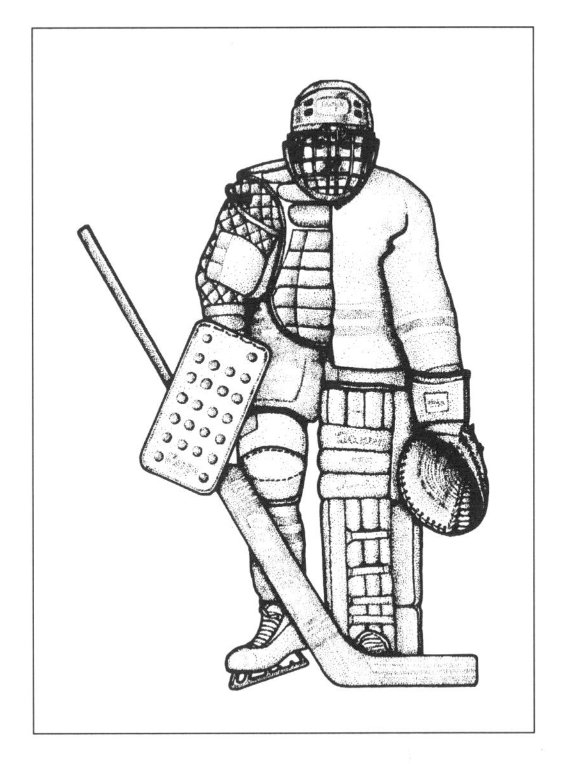

**FIGURE 28**
PROTECTIVE SPORTSWEAR

The role of developments in materials technologies in the evolution of specialised apparel, and the range of materials which may be used, is demonstrated by this goalkeeper's outfit for ice hockey which is designed to give full protection against all the impact hazards associated with the goalkeeper's role. The helmet incorporates a face mask made from steel coated with epoxy resin and a high-density polycarbonate throat protector. The shoulder and arm pads are made from nylon-covered foam and are designed to combine a high degree of mobility with protection. The body pad is made from Rubatex foam, covered and lined with waterproof nylon. The goal pads incorporate foam padding and a cow hide covering. The stick hand mitt has a sheet of polyethylene sandwiched between polyfoam and leather. Other specialised materials are used elsewhere.

**SAQ 8**
The last paragraph links back to examples in earlier Blocks, suggesting some sort of pattern may be evident here. What might this be?

In all the examples above, the standards of performance which can be specified have been improving, sometimes dramatically so. To understand the sources of this improvement we need to look at the way that production and technological environments may shape textile design – and at how changes in the underlying technologies are linked to the factors that have been considered in this section. These are the main concerns of Sections 3 and 4.

# 3 TEXTILES EXERCISE, PART 2

In Section 4, I will go on to look at the manufacturing and technological factors which shape the development and design of textile products. But as a preliminary to this, there are two further elements of the exercise as a conclusion to the work you undertook in Section 1.

First, I want you to think further about the suggestion (in the Section 1 Introduction) that the textile sector is in some respects 'anachronistic'. I suggested earlier that the sense of anachronism is related to the continued use of the 'pre-industrial' **natural fibres** such as cotton and wool even though **manufactured fibres** of different types and varieties have become increasingly available, particularly in the last fifty years or so. During that time, the processes of incremental innovation have produced substantial cumulative improvements in the cost of manufactured fibres and in their properties.

Although the image of 1950s drip-dry shirts may persist as a reminder of the shortcomings of the early synthetics, some high-quality fabrics can now be produced from manufactured fibres alone. Their properties of strength, lightness, elasticity and so on surpass those of earlier generations of natural or manufactured fibres. From the point of view of manufacturers further down the supply chain, the manufactured fibres can be considerably easier to process, not least because they arrive at the spinning mill in a pristine condition and thus require less processing. They can be produced to any required length, either as a staple (short length), or as a continuous filament. Similarly, they can be produced in a wide range of thicknesses and with varied cross-sectional profiles which are matched to particular types of end-use – more generally, their properties can be specified to precise limits. As you have seen, precision in specification is a vital part of product development. The ability to form filaments of specified length, width and profile appears to offer the potential for by-passing the subsequent stage in production – that of spinning. Yet manufactured fibres are mostly produced in staple form, much of it for production into yarn on what is known as 'the cotton system'.

In contrast, the natural fibres present a number of difficulties, bearing out the point that we are 'obliged to take most materials as we find them with all their natural properties, and to improvise things out of what we find' (Block 5, Section 1). All the natural fibres are randomly variable, adding to the tasks of processing in that they have to be graded, although this is only within comparatively broad parameters. The costs and tasks of processing are added to by their arrival in a condition which is generally dirty, includes seeds and other waste in the case of cotton and, in the case of wool, is greasy. Yet cotton and wool still account for about 40% of fibre consumption in western Europe and more than 50% of world consumption. While this represents a lower share than, say, thirty years ago, the output of cotton has continually increased – cotton crops are now thought to occupy about 5% of the world's cultivated land. Given their disadvantages, why have the traditional fibres retained such a large market share? Does this persistence provide an example of the suggestion made in Block 5 that people are sometimes inclined to have an antipathy for synthetic materials, and to favour those they think of as 'natural'?

In the continuation of the Textiles Exercise (below), I will ask you to examine the materials, components and construction of a jacket – a product in which there are marked similarities between contemporary methods of construction and those of a hundred years earlier. To the outsider at least, both the comparative simplicity of an item such as a jacket, and its apparently limited evolution, are two of a number of 'puzzles' which may strike an outsider looking with fresh eyes at the sector. Another puzzle, particularly for those familiar with engineering, is why it is that, seemingly after 17,000 years, sewing is still used to assemble most garments. In some modern clothing factories, the equipment for garment assembly includes machines of considerable complexity such as the one in Figure 29. As in the figure, these may be controlled by micro-electronics, and incorporate an array of sensing devices together with pneumatic, electro-mechanical and other ancillary equipment. Yet these sophisticated machines – the machine tools of the apparel industry – are still generally dependent on direct control by human operators who also have to undertake parts loading and are an integral part of the main machining function, skilfully guiding components through to achieve the correct profile in joining. The focal point of operation in all this remains the needle, albeit one that is power-operated. Why are more efficient systems for assembly – such as welding or bonding in some other way – not used, as they are for textiles used in the industrial sector? Related to this is one of the questions raised in Section 2: Why is it that the development cycles are so long?

These puzzles are just something to think about for the time being, and they will be taken up in Section 4. First, the conclusion of the exercise started in Section 1.

**FIGURE 29**

SEWING UNIT FOR PRODUCING POCKET OPENINGS OF VARIOUS TYPES

The machine incorporates a micro-electronic control system which enables rapid changes between styles, automated sewing along varied profiles, and automatic unloading and loading elements – although some elements of manual loading and unloading remain.

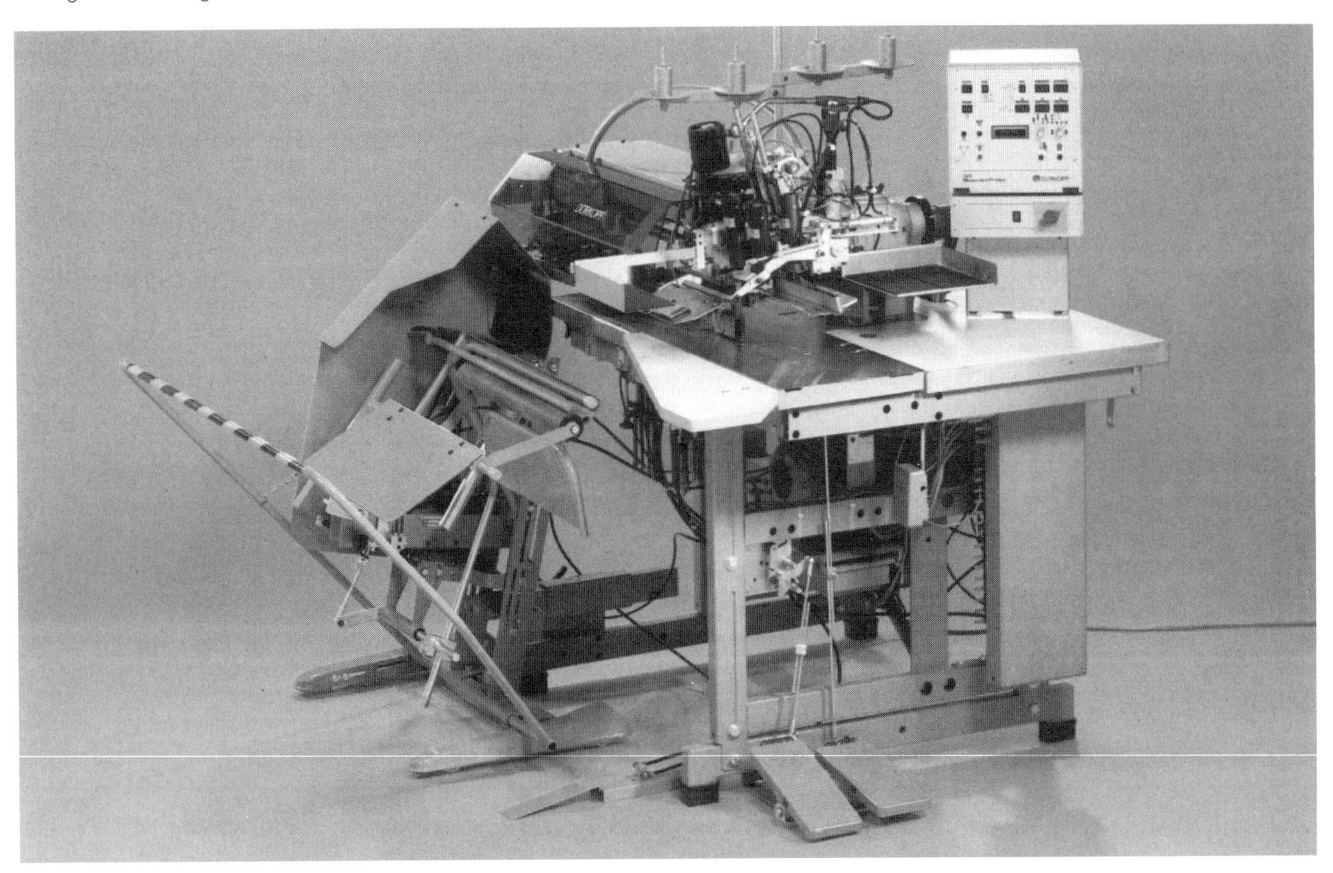

## TEXTILES EXERCISE, PART 2

With your work on the first part of the exercise and on Section 2 in mind, four further questions are set out below. Questions 7 and 10 relate to the broad range of products which you found in the first part of the exercise. For Question 8, I want you to select and examine a single item: a jacket – men's or women's will do – but preferably of the type that might be worn on formal occasions.

As a guideline, I suggest that you spend about an hour on this part of the exercise: 20 minutes or so for Question 7, and about 10 minutes each for the other three questions. Note down your answers.

**Question 7**

Bearing in mind the points raised in Section 2, in what respects does the list of performance requirements from BS 7000 contained in Block 2, Section 10.3 (p.105) seem to be relevant in developing design specifications for these products? Think about the applicability of each of the items in that list to one or more textile products.

The list from BS 7000 was as follows:

1 appearance and texture;
2 static requirements, e.g. size, mass and colour;
3 dynamic requirements, e.g. input and/or output;
4 ease of use;
5 environmental conditions of use, e.g. temperature, humidity and shock;
6 safety;
7 relevant standards and current legislation;
8 reliability;
9 maintainability;
10 disposability.

**Question 8**

Carefully inspect the jacket you have selected (referred to above). How does this appear to have been constructed? What – if any – manufacturing difficulties do you think there might be with a product of this type?

**Question 9**

Having examined your selection of half a dozen products:

(a) What types of materials have been used in their manufacture or construction? (The product labels or those containing cleaning instructions generally contain this information; otherwise, take a guess. A basic distinction you will find is between the natural fibres such as cotton, and manufactured fibres such as rayon and nylon.)

(b) How would you classify them within the materials typology suggested in Block 5: metals, ceramics, polymers, composites?

(c) What factors, do you think, might explain the balance between the use of manufactured and natural fibres?

**Question 10**

It has also been evident in the earlier Blocks that technological innovation does not take place in isolation, but that it interacts with a range of other factors, for instance, in the technology-push / market-pull model discussed in Block 3. Thinking forward now, in what ways do you think that the development and design of textile products might change in the next ten years or so?

**Complete this exercise before reading on.**

# 4 A 'DESIGN CHAIN'?

The structure of this section will follow from the questions in part 2 of the Textiles Exercise (Section 3). In Section 4.1, I discuss Question 7. This provides a number of important 'establishing' points against which the technical problems of textile design and production can more readily be viewed. In Section 4.2, Question 8 provides the starting point for a consideration of how product development interacts with manufacturing in one specific area, that of garment manufacture, and this is considered more fully in Section 4.3. Consideration of Questions 9 and 10 provides entry points for examination of the longer-term processes of technological innovation which lie beneath the more ephemeral concerns with styling changes that were considered in Section 2. Two areas have been selected for their significance for product development: innovation in fibres in Section 4.4, and the interaction of innovation and design across the textile chain as a whole in Section 4.5.

## 4.1 SPECIFICATION AND PERFORMANCE

In general terms, all the requirements from BS 7000 (Question 7) are relevant for textile products – although they apply with varying significance. When considered in aggregate against specific items, particularly clothing, it is clear that there is a substantial range of pre-conditions which have to be considered in the specification. Looking at the individual requirements of BS 7000 as listed in the question:

1 I hope it will be clear from much of the content of Section 2 that considerations of appearance and texture are central. Indeed, in terms of the short product lives I referred to in Section 2.3, it is this element of specification which is most likely to drive product design in the short to medium term from one season to another. But not in the longer term: over time, more fundamental changes in specification may be possible, often through changes in product or production technologies at any of the stages in the production chain.

2 Of the types of static requirements listed, the role of colour should also be evident from Section 2. However, size and mass raise some particular problems in textiles, both in measurement and in production. These are considered later.

3 At first sight, dynamic requirements appear to have little, if any relevance, particularly when thought about in the more usual input/output terms associated with, say, mechanical and electrical equipment. But there are inputs and outputs in a broader sense, such as in the interactions between the human body, clothing and the surrounding physical environment. In one sense, clothing acts as an intervening variable. For instance, it plays an important role in temperature regulation, protecting the body from the cold and from overheating, and this role is one of the factors which shapes product specification. But the body also produces outputs – of heat and of the slightly acidic moisture of perspiration – which interact with textiles, and which are mediated by them. People's feelings of comfort in wearing particular clothes are directly related to the efficiency with which the moisture is dispersed – a factor which, in turn, is directly related to the varying properties of fibres (for instance, cotton is more effective than most manufactured fibres), and in the retention or dispersal of body heat. These considerations can affect the styling of clothes, but are particularly important in the selection of fabrics – of different weaves or knits and, as part of this, of yarn and fibre types.

4 Other than for some highly specific occupationally related circumstances – such as the design of the space suits in Figures 30 and 31, this requirement appears to be the least problematic. Questions of ease of use have either been long solved or are consciously over-ridden by considerations of fashion and style (Figure 32).

**FIGURE 30**
SUIT DEVELOPED FOR THE APOLLO SPACE FLIGHTS IN THE LATE 1960S

The spae suit was worn over a liquid-cooled undergarment and, as can be seen in (B), was of a multi-ply construction which incorporated The following layers:

1 – a layer of heat-resistant Nomex (a nylon-type polyamid fibre with a high melting point of 378 °C, which is also used in protective suits for racing car drivers);

2 – a gas-tight bladder of neoprene-coated nylon;

3 – a containing layer made from nylon;

4 – a further neoprene/nylon layer to provide thermal and micrometeoroid protection;

5, 7, 9, 11, 13 – alternating layers of perforated aluminised Mylar (non-textile PET – polyethylene terephthalate – film);

6, 8, 10, 12 – alternating layers of a non-woven polyester fibre – Dacron;

14, 15 – two layers formed from a polyamid fibre with a high melting temperature – Kapton;

16 – layer of 'beta' cloth;

17 – outer layer of Teflon fabric.

(A)

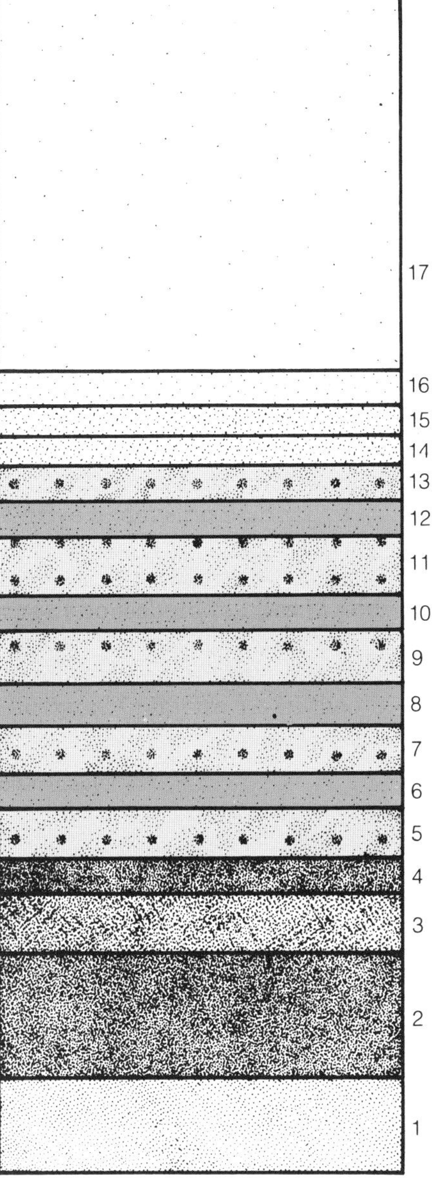


(B)

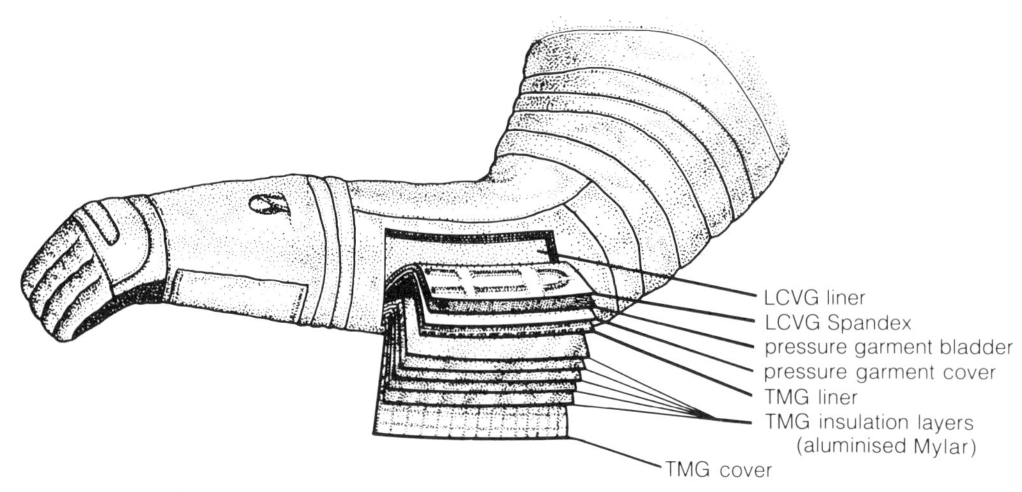


**FIGURE 31**

SECTION FROM THE SUITS USED BY THE CREWS OF SPACE SHUTTLES

In contrast to the Apollo suits, which were custom-made for each astronaut, these suits are comprised of separate components which can be assembled to fit most potential male or female crew members. A typical cross-section is of 11 layers consisting of a Liquid-Cooled Ventilation Garment (LCVG) – two layers – to maintain astronaut comfort; a pressure-retention garment – two layers – to provide containment of the breathing air; a Thermal Micrometeoroid Garment (TMG) – 7 layers – to insulate against temperature extremes and protect against micrometeoroids.

**FIGURE 32**

SKATER

This fashion concept, largely using non-textiles, uses methods recognisable from Block 5. The construction is of closed-cell foam which is laminated and profiled by vacuum moulding. Influences from the protective sportswear in Figure 28 are detectable.

5 Across the whole spectrum of textile products, a wide range of environmental considerations shape the choice of materials and fabrics – for clothing and for other products. Some of these were referred to in Section 2.7, such as resistance to chemicals and sunlight. In addition, clothing is often worn in circumstances which are dirty, for instance where body and clothes come into contact with a range of air-borne particles generated by road and other traffic, or from other sources of air pollution. Garments need the capacity to resist these and to allow their removal during cleaning without loss of garment integrity.

6 The issue of safety has two dimensions. The first concerns identification and avoidance of the hazards which clothing and other products may incorporate. The most obvious hazards concern small children, such as those associated with burns where clothing catches fire, and there are other fire-related hazards in the home, for instance in upholstery and other furnishings. Hence, there are specific British Standards covering both these possibilities. Hazards also arise in the workplace, for instance, where loose-fitting clothing may be caught in machinery. But, as was pointed out in Section 2.7, textile products also have a role in providing protection (for instance, fire blankets and heat-proof gloves).

7 Legislation and standards partly arise from the points in 6 above but, since textile processing often involves various chemical treatments, these also need to be considered.

8,9 Reliability and durability were also briefly referred to in Section 2.7. In essence, end-products generally need to withstand considerable wear and tear – particularly in children's clothing – in day-to-day use. Garments need to be *engineered* to withstand these physical demands at an acceptable level. All present problems in their cleaning. In part, these are problems of maintaining characteristics of colour, shape and so on, but it is also important to take account of the impact of the various cleaning processes on the environment. This is taken up in Section 6.

10 The disposal of textiles can present a number of problems. Synthetic fibres can contribute to the generation of toxic gases as their chemical structure breaks down if they are incinerated, or after disposal in landfill sites. In a different context, growing use of textiles in cars (together with plastics) contributes to problems of scrap recovery and to problems of capacity in landfill sites (see Section 6).

## 4.2 INTERDEPENDENCES IN MANUFACTURING

In Question 8, I asked you to look at the construction of a jacket, and to envisage what difficulties there might be in its manufacture. What you are likely to have found is that it was constructed from a range of different types of fabric – most obvious in the difference between the inner and outer body of the jacket – and that these fabrics were made of various types of fibre – cotton, polyester, and so on. Although it may seem unusual to think about it in this way, it is similar to most other manufactured products in that it is an assembly of components which has been engineered to meet specific requirements and dynamics of use (stresses, etc.). In the same way that a car is assembled by first producing a series of sub-assemblies which are then combined in final assembly, so an item such as a jacket is also produced via a series of sub-assemblies. Of course, there is an enormous difference in the magnitude and the complexities of the tasks involved, but in both cases the rigour with which design is undertaken determines commercial success, albeit in rather different ways. To establish how this is so, I will first look at the sample jacket of Question 8.

First, the components and the materials. In the case of the jacket I looked at, surmising about the nature of the hidden items, I compiled Table 1 to show the various components.

**Table 1 Components of a jacket**

| Outer body (woven, wool) | Other components |
|---|---|
| Back – 2 panels | Body linings – 19 panels (viscose – rayon) |
| Front – 4 panels | Pocket edgings – 8 (viscose) |
| Sleeves – 4 panels | Pockets – 12 panels, synthetic fibre (polyamid?) |
| Lapels – 2 panels | Pocket edge pipings – 4 |
| Collar – 1 panel | Interlinings – 5 components (probably synthetic fibre – polyamid?) |
| Pocket flaps – 3 | Shoulder pads – 2 (polyamid?) |
| | Collar lining – 1 (wool) |
| | Coat hook tab – 1 |
| | Labels – 3 |
| | Buttons – 7 large, 7 small, plastic (compression moulded?) |
| | Sewing threads – various for button attachment, button holes, seams, darts, etc. (counted as one item) |
| Total components, outer body – 16 | Total, other components – 70 |

My total of 86 components is an underestimate, certainly in one, and possibly in two respects. At the component level considered above, there may be another half dozen or so more components in the interlinings (which are used to reinforce some areas and which contribute to the retention of garment shape). More importantly, it should be kept in mind that the components listed above are themselves assemblies from the prior stages of fibre production, spinning and weaving. As will be seen in Section 4.4, it is these stages which provide the longer-term impetus of innovation and the development of new products.

The number of components in the example above is larger than might be expected – particularly in the number of 'other' components. In all discernible cases, down to the labels, the various components will have been assembled by machine stitching, although bonding using heat-setting adhesives might have been used for some of the inner linings. Sub-assemblies will have been used for the main lining, side panels, and so on. Assembly will also have involved the insertion of various sewn darts and tucks which, like the interlinings, are used to give shape to the jacket. Since the fabric is a fairly complex weave which uses yarns of five different colours in a wide-checked pattern, considerable care has been taken with the cutting of the outer components to ensure that, when they are aligned in assembly, the continuity of the weave patterning is maintained in the garment as a whole.

The jacket I looked at, as you will probably have surmised, was a fairly expensive item, and towards the upper end of the range of complexity for everyday outer clothing. Many garments contain considerably fewer components, and are likely to present fewer complexities in assembly. You may have thus concluded that, even if considerable skill in design and manufacture is evident, by comparison with most other products, it is unlikely that there are any unusual manufacturing difficulties. But only in some respects would this conclusion be correct. So, what are the difficulties, and what can be learned from them? This question is taken up in Section 4.3.

## 4.3 DETAIL DESIGN

In Section 2, it was mentioned that many of the problems faced in garment design for large-scale markets arise from the need to reconcile the styles or designs for particular garments and for garment ranges with the demands of the manufacturing process, and with the economic pressures that are faced because of the intensity of competitive pressures and from the ready availability of low-cost products, such as from manufacture in the developing countries. As you saw in Figure 17, this process of reconciling styling and other design considerations can involve an elaborate, multi-stage process, with much iteration and involvement of various groups concerned with design, marketing, manufacturing and so on. Unless they can be resolved in the early stages (which is often not the case), a considerable number of these difficulties ultimately focus on the detail design stage, and I will concentrate on these in this section. Among other things, this provides a useful contrast to the problems of design for manufacture which were considered in Block 5, and it re-iterates the vital role in design of specific **domain technological knowledge**.

### THE ROLE OF DETAIL DESIGN

In Section 2, it was suggested that, in the short term at least, fashion design – or styling – plays a more important role in the overall process of textile product development than is generally the case with other types of product. As you saw, the process of exploring options in such respects as garment shape, fabric colour and surface design, and the narrowing of the options to the point of commitment to a specific product or product range, correspond to the first three of the design stages discussed in Block 2, Section 12.2. Cost estimates often cannot be based on detailed investigation because of the time pressures of the design-to-production cycle for products manufactured in large volumes, and because there are generally a large number of competing design concepts until the final stages of the decision process narrow the options down. Past experience provides a basis for estimates of the amount of fabric required for each garment, of the labour costs involved, and of other costs. Particularly where production is undertaken as a contract for a retailer, and/or an unfamiliar type of fabric is involved, there is a considerable potential – which sometimes becomes actual – for there to be a loss on manufacture. The tasks which equate to the detail design stage are critical in avoiding this.

> What activities would you expect to be involved in detail design, and how might they be critical for success? (Think about this for a few minutes.)

---

> This is discussed in the following text.

**FIGURE 33**
MAIN ACTIVITIES IN DETAIL DESIGN AND MANUFACTURE OF GARMENTS

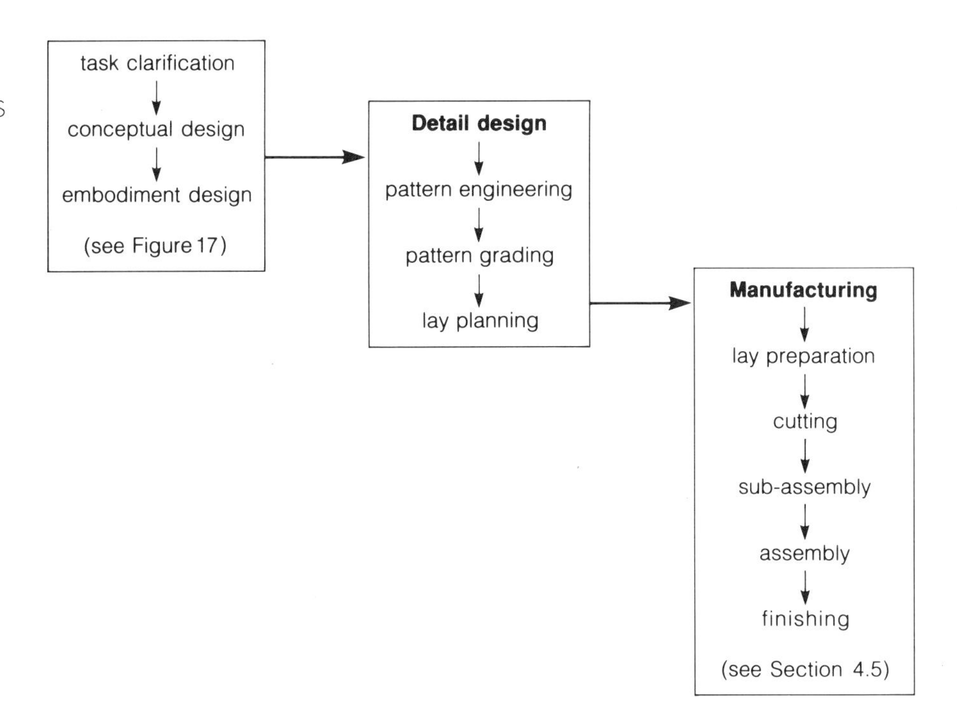


The main sets of tasks involved in detail design are outlined in Figure 33 (together with the main stages in manufacturing). However, an important preliminary point is that the activities listed in the figure lie at an uneasy point in the interface of design with manufacturing. A key part of the tasks involved is to attempt to reconcile the intentions of the fashion designer with the constraints of manufacture. This is sometimes in the practical sense of what is possible, but more generally in terms of what, at least, is financially viable. The activities indicated as detail design are sometimes geographically separated from the conceptual and other earlier stages of design, being located within the manufacturing unit. Thus, they may not be acknowledged as part of the design process by many 'clothing designers', and the changes that may be made in pursuit of producibility can be deeply resented. The issues of **design for manufacture** which you considered in Block 5 are thus a very live focus of concern – although in rather different ways. For some specific products, it has proved to be increasingly attractive in competitive terms to use some of the 'design for manufacture' (DFM) principles – or, good practice – which were discussed in Block 5. For instance, **modular design** and use of standard components do allow an increasing element of **customisation** where a product as simple as a shirt becomes the basis for a **family of products** varying on relatively minor lines such as fabric print, pocket attachment and cuff or collar style. But there are more fundamental, general, and prior notions of DFM which are important in ensuring viability. These are considered below.

**SAQ 9**

How would you summarise the tasks involved in detail design?

Using the three-tiered classification of processes in Block 5, garment manufacture involves only **secondary processing** and **assembly**. **Primary processing** is the concern of the 'upstream' industries in the textile chain – fibre producing, dyeing and finishing. (These are considered in Section 4.5 together with the other assembly processes of yarn spinning and weaving.) Knitwear production combines one element of primary processing with secondary processing and with assembly. In all three process categories, the nature of textile materials is such that some unusual types of processing problems are involved. In Block 5, the examples were all of products which would hold a defined shape after a given process. With textiles, this is generally not the case. Throughout the primary, secondary and final processes of manufacture, materials and parts are generally highly flexible in character, and do not in themselves retain a given shape – in any dimension. This becomes particularly important when the stage of garment manufacture is reached.

## FABRICS AS STRUCTURES

In Block 5, you looked at the requirement to be able to select materials for particular tasks on the basis of awareness of their properties. How does this apply to the example of textiles and, more specifically, to garments? (Think about this for a few minutes.)

---

For textiles in general, the starting point is that the various types of manufactured and natural fibres all have different sets of properties – see Section 4.5 below. These properties are added to by the processes of assembling fibres into spun yarns of different types and by the formation of yarns into fabrics. In addition, various types of chemical and other processing modify the properties of fibres, yarns and fabrics, for instance, to add colour, reduce shrinkage, and so on.

Taking the point at which fabric properties have to be considered for the design of everyday garments, what technical properties may be relevant? (Think about this for a few minutes on the basis of the technical properties of materials identified in Block 5 and, if this is helpful, look again at one or two of the fabrics identified in the first part of the exercise.)

---

The importance and role of such 'domain knowledge' for design is discussed in the following text.

If engineering properties appear to be far removed from textile products, you need to think back to the materials samples from the kit you worked with in Block 5. Carbon fibre and Kevlar are examples of textile materials which are displacing 'traditional' engineering materials for some applications. Although those fibres are the product of development processes aimed at very different types of specification, they nonetheless have affinities with other types of textile. *Low density* is a desired property related to end-use and, in varying magnitudes, this needs to be combined with a significant level of *strength*. As you saw in the example of carbon fibres, strength is partly derived from the characteristics of the fibres themselves, but it is also a product of the way that fibres are assembled into load-bearing *structures*. In textiles used for clothing, these structures are formed in two stages – in yarn spinning (where the fibres are aligned along the direction of the yarn), and in fabric formation.

In the case of woven fabrics, the structural form is that of a matrix consisting of warp yarns (along the length of the fabric) and weft yarns (across the width) as can be seen in Figure 34. These are *tensile* structures in that the forces they have to resist are 'pulling' forces – activity by the wearer of a garment can result in considerable tensile force. Most fabrics have the capacity to resist these forces by allowing

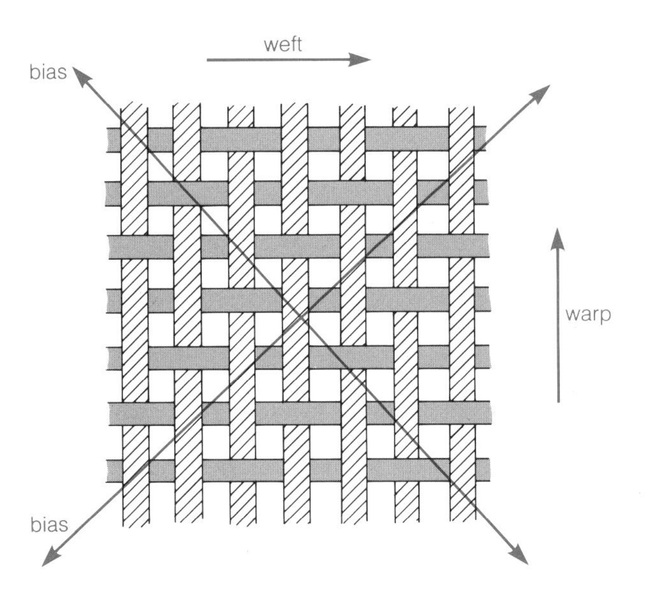


**FIGURE 34**
WARP, WEFT, AND BIAS

some measure of distortion while retaining the capacity to recover their original shape – i.e. they have the property of extensibility, or *elasticity*. Of the other properties identified in Block 5, it is very low levels of compressive strength and hardness that are generally desired. The first of these is necessary because low compressive strength is correlated with a low shear modulus, a property which is manifested in fabric 'drape' – the ability to mould to the shape of the body. Thus, the propensity of silk fabrics to readily adopt the contours of the wearer is a product of low shear rigidity. Finally, low levels of hardness – mainly resistance to abrasion in this context – are, of course, one of the more important factors in wearer comfort.

But textile structures are complex in that they are *anisotropic* – they respond in different ways to tension on different planes. If you examine a (non-elasticised) woven textile and put it under tension by pulling along the lines of maximum tensile strength (the line of either weft or warp), you will usually find that there is only very limited extensibility, although there are differences between the warp and weft directions in this respect. But if you pull on the line of the diagonal of the weft and warp, the degree of extensibility is somewhat greater – although the extent of this variation depends upon the type of yarns used in the weave, and on the weave density. These variations in extensibility limit the ways in which a fabric can be cut. Fabric cutting, and thus component design, has to ensure the retention of shape in a garment by aligning the weft and/or warp directions to correspond with the main directions of stress in any given component (see Figure 34). The problems are also significant in assembly – think of how seams in jacket shoulders (for example) are sewn in an arc across the weft, warp and bias directions. Considerable skill and adequate allowances of fabric for seams are necessary to overcome this. Hence, one of the tests of garment quality is to look at the finish on this seam.

However, there are considerable differences in properties between fabrics. These are most obvious between different categories of fabric. For example, stretch fabrics such as those using the trade name Lycra are different in that they include a special group of fibres with *elastomeric* properties. These allow a much greater element of extensibility across all the planes of a fabric. Similarly, knitted fabrics have generally high levels of extensibility. In this case, extensibility derives from their structure of intermeshed loops as can be seen in the example in Figure 12(C).

The ways in which different fabrics behave as structures, and associated variations in strength and extensibility, are some of the properties which garment designers have to take into account, and sometimes exploit, to produce particular types of fit, hang or shape in a garment and to design it to withstand the rigours of use and of maintenance. But these properties are also a source of difficulty in both design and manufacture. One problem is that fashion designers do not always possess the skills and knowledge for garment engineering – to ensure that the issues of manufacturability and of durability in use are addressed adequately.

Underlying many of the engineering difficulties is the problem that the forming processes of spinning and weaving create structures which have a number of in-built instabilities. These arise partly from the properties of the fibres themselves. For example, wool and cotton fibres are *hydrophilic* (i.e. they absorb water). This property is one of the reasons why they are valued for textile use – it is an important factor in a wearer's feelings of comfort because moisture is rapidly removed from the skin's surface. However, in manufacture, this also means that many fabrics are prone to dimensional instability – they do not retain precise dimensions if there are variations in humidity, or when they are processed in water-based solutions, such as some dyes. The potential for instability is increased by any residual tensions or stresses that may remain within a fabric from the spinning and weaving processes since they may manifest themselves during manufacturing – for instance, cutting fabric may release some of the tensions, and a cut part will thus change shape. This contributes to a number of problems and added costs in handling and processing. For instance, because the dimensions of fabric can alter significantly with changes in humidity and temperature, weavers starch some fabrics to maintain stability. The cost of this is increased by the need to remove the starch in later stages of processing. (This is an example of the conflict between the properties of a material for *performance* and its properties for *processing* identified in Block 5.)

Thus, cutting components and assembling garments with precision can be difficult. To avoid the problems of instability requires a variety of constructional approaches, stabilising treatments, controlled environments and so on. More fundamentally, the porous and limp nature of most clothing fabrics has, until recently, constituted a barrier to the mechanisation and automation of both component handling and most individual manufacturing processes. Among other things, this means that a high proportion of non-materials costs are concentrated in the area of garment assembly. This is because assembly is largely reliant on machining operations in which component handling and the skilled manipulation of components in the joining processes are the main sources of cost. For instance, Hoffman & Rush (1985) estimate that labour costs for machinists amount, on average, to about 80% of total labour costs, and between 30 and 40% of total costs. But much of the machinists' time – more than 80% is the general industry estimate – is spent on fabric *handling* rather than machining. This concentration of costs has focused the search for cost reduction on the automation of machining, but progress has been slow in this respect. Fabric handling presents great difficulties. For instance, to approach the accuracy and adaptability of a skilled machinist, successful robotic application in most areas would depend upon the use of some highly expensive and, as yet, unperfected combination of vision and sensing devices combined with high-precision manipulation, and it would need very sophisticated software control systems.

## THE NEED FOR OBJECTIVE MEASUREMENT

The structural characteristics of fabrics, and their instability, contribute to a further set of problems – those involved in finding objective systems of measurement. Until recently, assessment of the characteristics and properties of different types of textiles has largely been informal and essentially intuitive and subjective in character. For instance, colour was assessed by eye, while the tactility, drape and other characteristics of fabrics have been primarily assessed in terms of 'handle'. This assesses fabric by touching, pulling, crushing, draping and otherwise manipulating it to provide an evaluation which is based upon accumulated craft skills and experience. As a 'measure' it is important both for characteristics in wear and for the manufacturability of a fabric. Some impression of the subjectivity involved is provided by Table 2, which represents a codification of what Japanese tailors understand by 'handle'.

**Table 2 The 'hand' or 'handle' of a fabric, as expressed by Japanese tailors**

| Japanese term | English term | Definition |
|---|---|---|
| *Koshi* | Stiffness | A feeling related to bending stiffness. Springy property promotes this feeling. Fabric having a compact weaving density and woven using springy and elastic yarn gives this feeling strongly. |
| *Numeri* | Smoothness | A mixed feeling: limber and soft. Fabric woven from cashmere fibre gives this feeling strongly. |
| *Fukurami* | Fullness and softness | A bulky, rich and well formed feeling. Springy property in compression and thickness accompanied with warm feeling is closely related to this feeling. (*Fukurami* means 'swelling'.) |
| *Shari* | Crispness | A feeling which comes from crisp and rough surface of fabric. This feeling is brought by hard and strongly twisted yarn. This brings us a cool feeling. (*Shari* means a crisp, dry and sharp sound arising when the fabric is rubbed with itself.) |
| *Hari* | Anti-drape stiffness | Anti-drape stiffness, no matter whether the fabric is springy or not. (*Hari* means 'spreading'.) |
| *Sofutosa* | Soft feeling | Soft feeling, a mixed feeling of bulky, flexible and smooth feelings. |
| *Kishimi* | Scrooping feeling | Scrooping feeling. A kind of silk fabric possesses this feeling strongly. |
| *Shinayakasa* | Flexibility with soft feeling | Soft, flexible and smooth feeling. |

*Source:* Sørenson (1990), derived from work by Kawabata.

What difficulties do you think might follow from a method of assessment like that described in the table?

---

The obvious difficulties are the imprecision and subjectivity of the terms employed, particularly for those outside the closed area of craft mystique. Even for those sharing the craft tradition, there can be difficulties in agreeing what constitutes a 'good', 'poor' or an acceptable handle – i.e., of quantifying and accurately

> *communicating* an assessment of handle – and of laying down a standard in a specification. Associated with these difficulties are a number of others, such as precision in the maintenance of quality standards, and of *replicating* these standards where production takes place at different factories or where there are repeat orders. This is becoming increasingly important as firms seek to move towards production in small batches, and to produce *repeat* orders in direct response to demand levels in the shops (see Section 4.6). Lack of precision can also contribute to the wider problems of cost control, for instance, where manufacturing problems follow from the choice of a particular fabric which proves to be difficult to process.

The codification of 'handle' referred to above was a starting point for investigation into the development of more precise means of assessment. It was realised that much of what is termed 'handle' can be related to a series of mechanical properties of fabric. These can be accurately measured and thus provide the basis for more objective assessment and comparison. Table 3 sets out parameters of measurement which have evolved from Kawabata's original work in this area. As the example of the surface and mechanical parameters identified in Table 3 show, these measures range from intuitively obvious measures such as fabric weight and thickness to more complex measures of structure and structural behaviour which have long been familiar in engineering. This methodological development is leading towards resolution of the types of problem such as accuracy in the specification of mechanical properties which were referred to above. It has also provided the basis for the dynamic simulation of fabric behaviour. Behind this lies a need for objective measurement and precise, consistently repeatable standards in fabric properties which are essential for further progress in the automation of production.

**Table 3 Fabric mechanical and surface parameters**

| Parameters | Description |
|---|---|
| Tensile | linearity of load/extension curve |
| | tensile energy |
| | tensile resilience |
| | extensibility, strain at a specified tensile load |
| Bending | bending rigidity |
| | hysteresis of bending moment |
| Shearing | shear stiffness |
| | hysteresis of shear force at 0.5° shear angle |
| | hysteresis of shear force at 5° shear angle |
| Compression | linearity of compression/thickness curve |
| | compressional energy |
| | compressional resilience |
| Surface | coefficient of friction |
| | mean deviation of coefficient of friction |
| | geometrical roughness |
| Thickness | fabric thickness |
| Weight | fabric weight (per unit area) |

*Source:* derived from Sørenson (1990)

Lack of objective standards of measurement is not a problem confined to garment manufacture; it is evident in other parts of the textile sector. The search for such standards and their (thus far, gradual) application are indicators of a more fundamental shift in the industries of the textile chain from a 'traditional' craft-based industry towards the more general pattern of industries which are able to draw upon a comprehensive, codified and organised knowledge base such as exists in engineering (Block 5). In this process, the development of objective standards of definition and assessment are an essential underpinning for the clarification of the 'problems' which provide the spur for new types of product, for understanding the problems, and for developing solutions.

### SAQ 10

What did Block 5 suggest were the components of the knowledge base of an industry or a professional group? What is the significance of the knowledge base?

## PATTERN ENGINEERING AND GRADING

I will now look at some of the specific tasks in the detail design stage. First, pattern engineering. Prototype garments are produced as a 'model' in a single, sample size. In some cases, these are the product of team-work between, say, a fashion designer, a textile designer, a skilled pattern maker, and a machinist specialising in the production of sample products. But in many cases, a prototype is developed by a fashion designer and machinist alone and, in these circumstances, less consideration may be given to questions of garment engineering and manufacturability. Although these issues are inevitably addressed in the decision-making process, this is at a general level. The key task of translating general decisions on styling, fabric type, product pricing and so on into designs that are *replicable* with precision and into profitable manufacture is that of the pattern maker – increasingly regarded as a pattern engineer.

The initial generation of patterns for prototype garments is primarily concerned with achieving the effect required by the fashion designer in terms of the shape of the garment, the drape and so on. In itself, this represents a considerable challenge, the nature of which is under-estimated through its familiarity. In essence, two-dimensional, non-rigid fabrics have to be converted into shapes that correspond to – and complement – the complex geometry of the human body.

### EXERCISE TAILORING TO SHAPE

If you doubt the complexity of the tasks involved, you might try a number of things. One is to attempt to clad a shape like, say, a cauliflower in newspaper or a similar 'fabric' in a way that results in a smooth 'tailored' appearance – *and* which minimises the waste of material.

You will find that, to achieve this, it is difficult to avoid cutting away a considerable amount of newspaper. In effect, overlapping components have to be devised, and these then need to be rejoined. In some places, darts or pleats will assist the process.

You might also look at a jacket sleeve and think about the specific problem of developing it. At first sight, a sleeve might appear to be cylindrical in shape – but it would look very odd if it was actually made in this way. A cross-section of a human arm is neither precisely circular, nor stable at many points (because it varies with movement and loading), nor uniform, because its geometry varies considerably along both the upper and lower arm. To achieve what most people would regard as an acceptable fit for this shape, and to allow for the full range of potential arm movements, the sleeve has to be cut and stitched along variable curves, along the line of the arm, but most obviously, at the join with the jacket body. Also, the components for the sleeve have to be cut from the fabric in the direction that is most compatible with the varying stresses in wear.

The ability to achieve a satisfactory fit derives from a combination of intuitive skills combined with a knowledge of how particular fabrics will behave when cut in particular ways. The designer also needs to know how the behaviour of the fabric is likely to be modified by the way a garment is structured in four respects:

- in the subdivision of a garment into components;
- by the insertion of darts, gatherings, pleats and tucks which add shape to a garment (see the example in Figure 35);
- by the inclusion of sub-surface components of different types, such as interlinings, to introduce an element of stability in shape; and
- by finishing treatments such as steam processing which, by the combined application of heat and moisture, release the internal stresses and produce a 'set' of the finished garment shape.

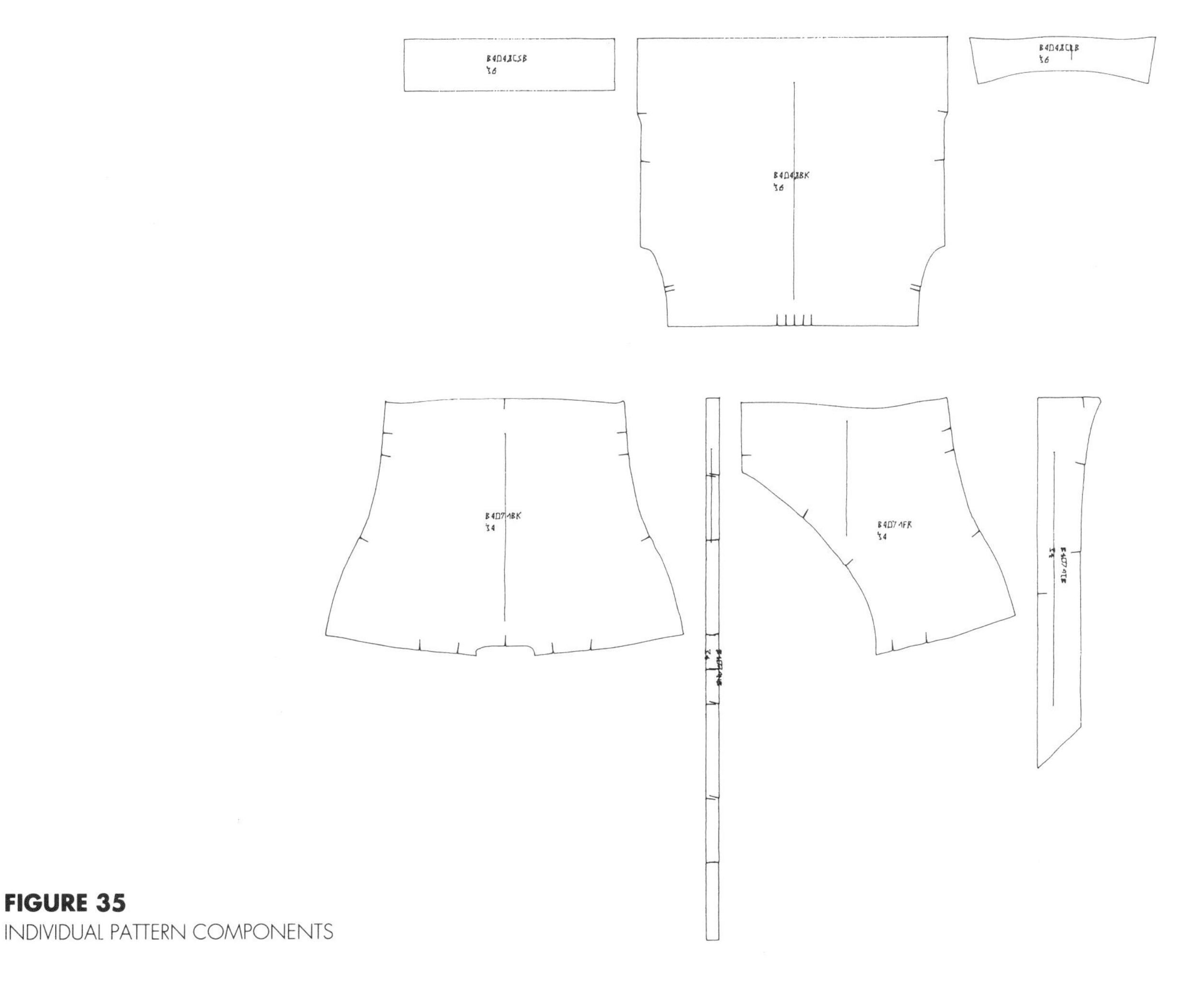

**FIGURE 35**
INDIVIDUAL PATTERN COMPONENTS

The skill of the pattern engineer lies in translating the prototype construction of a garment into a method of construction which retains the original design in terms of styling, which meets specified constructional standards of strength etc., while also establishing the most efficient approach to manufacture. This draws on a wide body of knowledge about fabric types and about methods of seaming and stitching (which are rather more varied in terms of approaches and contrasting properties than might be supposed). However, the pattern engineer may find that prior decisions on styling, fabric type, the quantity of fabric allowed per garment, estimates of labour costs, and target garment price cannot be viably accommodated in manufacture. Either the design will need to be modified significantly, or the cost and price parameters may have to change. In such cases, the immediate result is likely to be that the design has to go back through the decision-making and the fashion design stages. Not surprisingly, this possibility is the source of tensions between the manufacturing orientation of the pattern engineer and the styling and related concerns of the fashion designer.

In high-volume production for the mass market, the pattern engineer will have two particular, cost-related concerns in mind. The first of these parameters is **materials utilisation** since fabric costs alone are usually the major cost item – often amounting to around 50% of total costs. Thus the design of pattern components to achieve the minimum amount of fabric consistent with garment quality, strength, style and size, is critical. It should be added that the configuration of components for a particular garment is variable within the general parameters established by the style. Other decisions have to be taken on allowances for hems and seams. These partly depend on the type of fabric. In some cases, such as shirtings, fabric strength permits a comparatively narrow seaming allowance. In other fabrics, a small seaming allowance may result in a join that is likely to rupture in use. A further consideration is that too narrow allowance for seamings, hems and so on will result in either a high level of reject garments, or high labour inputs per garment, or a combination of these. There is a trade-off between materials (fabric) utilisation and **labour utilisation** since the second main concern of a pattern engineer is to ensure that handling costs and machining time in assembly are, so far as possible, minimised because this is the other main area of variable costs.

One of the differences between clothing and most other products is that every product type or variant has to be manufactured in several sizes. Thus, once a set of master patterns has been developed, patterns for all the various sizes in a range have to be generated from the master pattern shapes – see Figure 36. The total number of pattern shapes for any one product line can be considerable. For instance, the production of a 14-component main section of a jacket in six sizes may require, say, 84 different pattern shapes since only a few components, such as pocket flaps, can be used on more than one size of garment. Any adjustment in styling, or in the fabric used for its construction, potentially involves changes in most of the pattern shapes.

Because of the complexities of human shape, the process of pattern grading is not a straightforward one of proportional scaling up or down from the dimensions of the master pattern. For instance, arm lengths do not increase in a direct ratio with increases in chest size; people's body

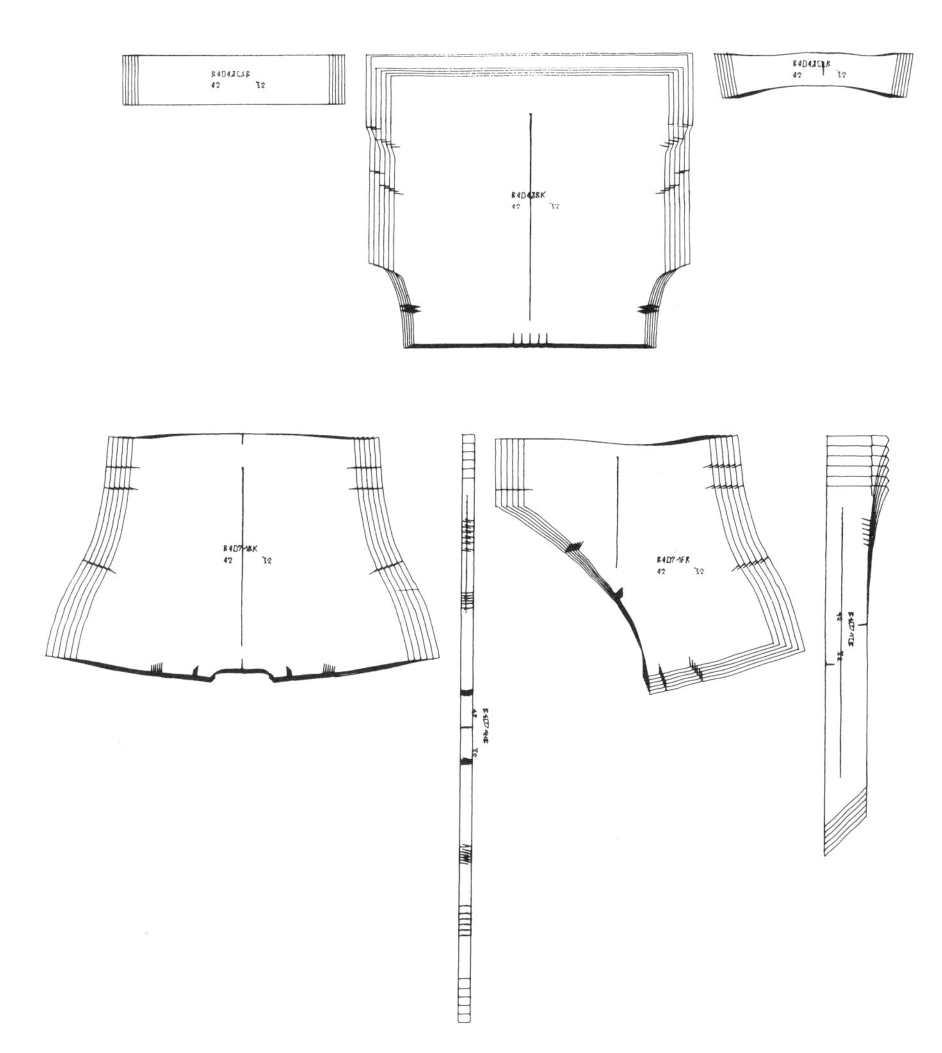

**FIGURE 36**
GRADED PATTERNS. NOTE VARYING GEOMETRY BETWEEN SIZES

shapes do not conform to regular geometrical shapes, but are irregular in contour, and curve along varying degrees of arc. There are British Standards for sizing, but for many retailers and manufacturers these represent a starting point rather than a definitive standard. As you will appreciate from the discussion of **user populations** in Block 2, any definition of garment size is necessarily a compromise, aimed at maximising applicability to a given section of the target population. But the nature of this compromise can vary between retailers according to the particular characteristics of their customer base. **Market positioning** by a retailer or for a branded product will draw a customer group which diverges from the general population distribution by socio-economic status, by age and so on. For this and other reasons, manufacturers or retailers tend to have their own rules for pattern grading – as you may have discovered by direct experience! To add to the complications, the grading rules have to be adjusted for different styles or cuts of the same category of garment.

## LAY PLANNING

The availability of computerised systems for grading has greatly reduced the time and the costs involved, and these savings are increased where the data generated in pattern grading are used in the next stage, that of lay planning. To maximise efficiency, woven fabric is usually laid up in multiple thicknesses for cutting, sometimes in thicknesses of 200 layers or more. Accuracy in laying up and in cutting are also fundamental in optimising fabric utilisation, and provide a stringent test for the effectiveness of one aspect of the pattern engineer's work. Success or failure in maximising the amount of fabric that can be cut from a 'lay' can determine whether a particular job is profitable or is a loss maker.

At the start of lay planning then, the task is to fit a complex range of shapes like those shown in Figure 35 into a rectangle of fabric (which may be 20 metres or more in length and, say, 1.5 metres wide) in a way that will minimise the amount of waste fabric.

> Which concept from the earlier course material helps explain how this process is approached?
>
> ---
>
> The answer, of course, lies in the concept of close packing which was introduced in Block 4, Section 2.3.

But lay planning presents more complex problems of packing than the examples you looked at there. In part, these problems result from the irregularities of the pattern shapes such as in Figure 35. The geometry of these shapes makes for a difficult fit within the rectangular shape of the fabric lay, as is shown in Figure 37. To some extent the problems of packing are reduced by the variable size of pattern components. For instance, the patterns for smaller components such as collar parts and pocket flaps can be inserted into the gaps between the larger components, as can also be seen in Figure 37.

A further problem arises in planning the cutting of the paired components used in a jacket, such as in the back, in the front, and in the sleeves.

> In these cases, is mirror symmetry involved? (I suggest you look again at the jacket you used for this part of the exercise.) What are the implications of this for planning fabric cutting?
>
> ---
>
> The answer is that mirror symmetry is often found, but not invariably, particularly in the front components which are most affected by styling considerations, such as the numbers and types of pockets, and so on. Where there is mirror symmetry, the implications for lay planning depend upon the characteristics of the fabric from which a garment is to be cut. If a fabric is two-sided, in that it has compatible face characteristics on both sides, then only a single pattern may be needed for paired components since a cut component can be used in either the right or left position. But many fabrics are not two-sided. In these circumstances, there are two possible solutions. Either paired pattern shapes can be used, or the fabric can be laid up in alternating face-up and face-down layers. However, the availability of the latter option depends upon the patterning of the fabric, and on its graining – some fabrics are 'one-way' in their surface grain or print design, and cannot be cut in this way.

Finally, another set of problems in lay planning are presented by the variable statistical distribution of the cut parts. This arises within a specific garment size, for instance in the difference between single and paired components, and in this case is simply resolved by halving the number of repeats of single components on the lay plan. But lay plans also have to accommodate the differences in the volume of components for the different garment sizes. Variations in the distribution of these components follow from differences in the forecast demand for particular sizes, and these may be in ratios that are less easy to accommodate.

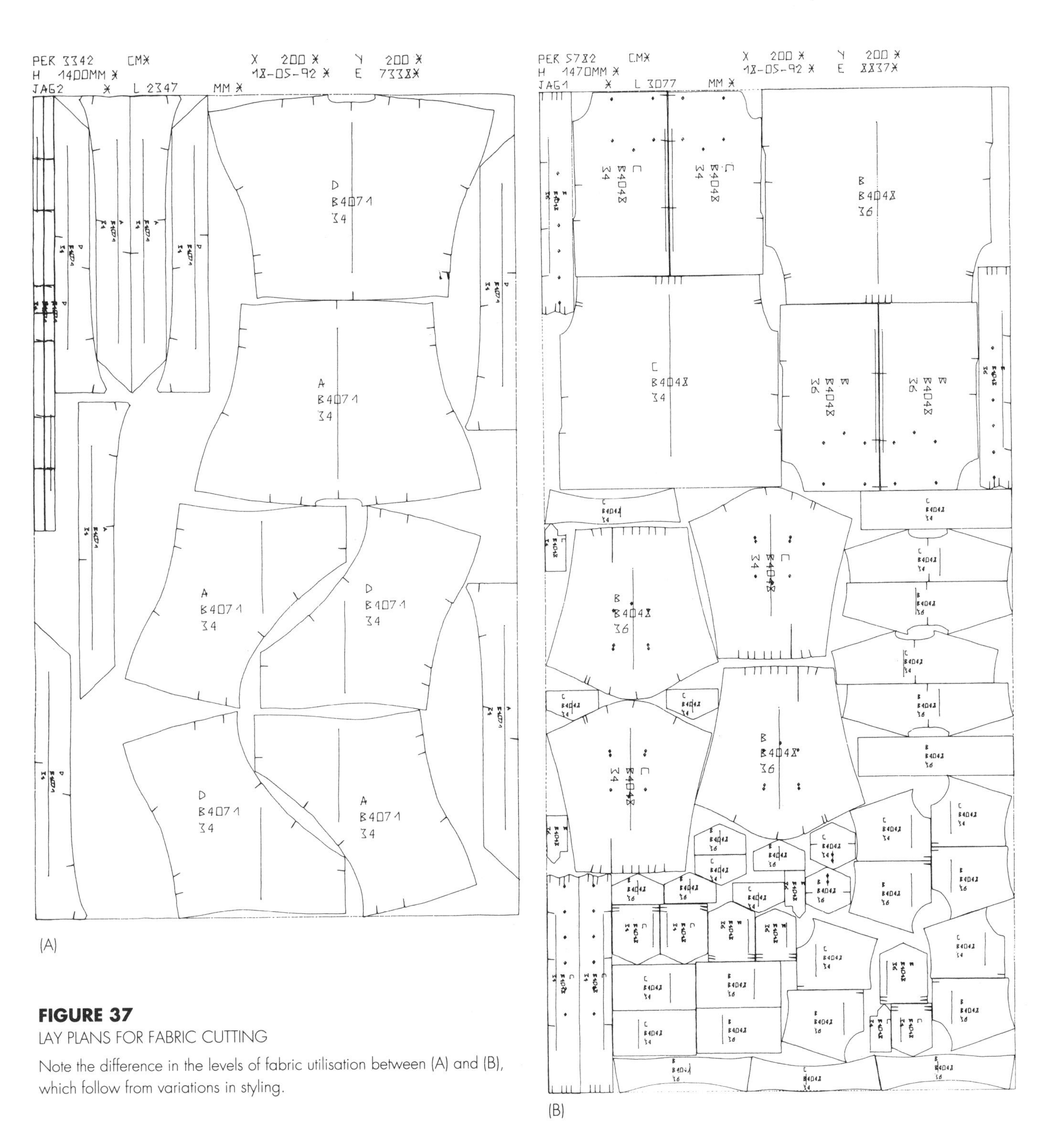


**FIGURE 37**
LAY PLANS FOR FABRIC CUTTING

Note the difference in the levels of fabric utilisation between (A) and (B), which follow from variations in styling.

The potential for failure to achieve satisfactory levels of fabric utilisation emphasises another important point from earlier parts of the course. This is that the flow of ideas, decisions and so on through the four design stages is not a one-way, downward flow, but an iterative process. I have already referred to the possibility that designs may have to be reconsidered as a consequence of work on pattern engineering. Similarly, if a satisfactory level of fabric utilisation cannot be achieved in lay planning, then it is necessary to look again at the master pattern design to investigate whether an alternative component configuration is possible.

For instance, it might be possible to move a seam position to change the *relative* size of components, or to divide one of the larger components in an area which is not generally seen. If this cannot be done, then the initial design may need to be modified in some way, for instance, by reducing the allowance for a pleat or some other stylistic feature. However, this risks compromising the integrity of the original design. In effect, a balance needs to be struck between the need to maximise the revenue from a design – which in great part depends upon the quality of the fabric and of cutting, together with the standard of assembly – and the need to contain fibre costs and labour costs.

The essential, general point is that the varied and often complex concerns of detail design have a central and integral role in ensuring the success of a design (measured on a number of dimensions besides cost). It is emphatically *not* a consequential, routine task, although it is sometimes regarded as such.

## 4.4 MANUFACTURING

The main stages in manufacturing were identified in Figure 33. Some of these are discussed briefly below.

### LAY PREPARATION AND CUTTING

As mentioned earlier, fabric is often cut in multiple layers, and this process carries through the emphasis on maximising fabric utilisation. Any error in preparing the fabric lay – such as in alignment or in maintaining the correct fabric tension – may result in cutting inefficiencies or in faults which are potentially replicable through the whole ply. In most factories, the next stage – cutting – remains a manual process, generally through the use of power-driven rotary or band knifes which use pattern shapes in a paper or card form as templates. As yet, automated, computer-driven cutting systems have not been very extensively used, although they have been available for some twenty years. Their high cost has generally limited their application to high-volume production. However, the expiry of a key patent has allowed the development of a range of competing new systems, and the economic volume for automated cutting is gradually being reduced to lower levels.

### EXERCISE A FABRIC CLAMP

Once computerised systems had been developed for pattern grading and for lay planning, it was comparatively easy to use the data from these stages in the control of an automated cutter such as a powered knife, or in a water jet or laser cutting system. The main problem was to develop a method for clamping the fabric securely in position to ensure accuracy in cutting. The system which was eventually successful was of elegant simplicity. As an exercise based on what I have said about the problems of handling textile fabrics, and drawing on some of the techniques for creative thinking discussed in Block 3, you might care to think about what methods might be used. The method which has been commercially successful derives from technical principles you have examined in an earlier part of the course. (The answer is included at the end of the SAQ answers section.)

In the larger, more capital-intensive factories, product development and the stages of manufacture through to cutting have changed radically in recent years. Both in organisational terms and in geographical location, they appear to be increasingly moving towards detachment from the downstream processes of assembly and finishing. For instance, many clothing firms in the European Community send cut parts to locations in eastern Europe and the Mediterranean region to be assembled. (There is a parallel for this in the printing industry where the adoption of computerised typesetting, and the use by customers of word processing and desk-top publishing systems, has led to a growing detachment of typesetting from printing and its integration with graphic design.)

## ASSEMBLY

The discussion of pattern grading and lay planning above re-iterates the emphasis in Block 5 on design for manufacture. But there is a further linkage of design and manufacture in that limitations in manufacturing capacity can restrict the options available for the development and successful production of new products. These limitations may be rooted in inadequate levels of investment in capital equipment or in inappropriate investment, for instance in equipment which limits the types of product design that can be undertaken. However, problems of this type are potentially capable of resolution in the medium term by the purchase of new equipment. A more intractable and long-term set of constraints is presented by shortcomings in the capacities of the design and manufacturing workforce as a whole. Lack of appropriate skills and expertise present many firms with serious difficulties as competitive conditions change. These difficulties are particularly significant in the area of garment assembly.

In contrast to the cutting and pre-cutting areas, garment assembly has tended to remain labour-intensive. Faced with highly competitive conditions, many garment manufacturers have responded by utilising new equipment where this has been available – and viable. There is a growing range of computer-controlled systems for moving parts between workplaces – although the parts still have to be loaded and unloaded by hand, and the systems are only viable in factories where levels of throughput are comparatively high. There is also a substantial range of equipment for the automation of specific assembly operations, for instance, the automated (but hand loaded and unloaded) insertion of button holes shown in Figure 38. Where costs are subject to severe, long-term downward pressures, even a seemingly minor development such as the semi-automation of label insertion can make an important contribution to sustaining viability. In addition, there has been emphasis on improving methods of work flow and other aspects of work organisation, particularly where this has helped to improve levels of labour utilisation. There has also been increased work intensification in the assembly area in which machinists may be confined to a single task which is undertaken at very high speed. This is thought to maximise productivity for the manufacturer, and it appears to offer higher earnings through the use of piecework payment systems. But the benefits – for firms and employees – may be illusory, and it is increasingly recognised that there can be serious adverse effects.

In this and other respects, the general pattern in the U.K. clothing industry appears to differ somewhat from that in countries such as Japan and Germany. As was pointed out in Section 2, much of the U.K. industry has moved to sub-contracting for retailers. Some of the implications of this are suggested by a comparative study of clothing factories in Britain and Germany (Steedman & Wagner, 1989). Comparing the manufacture of women's outerwear, it was found that the typical production run in British factories was in the region of 15,000 garments, whereas in the

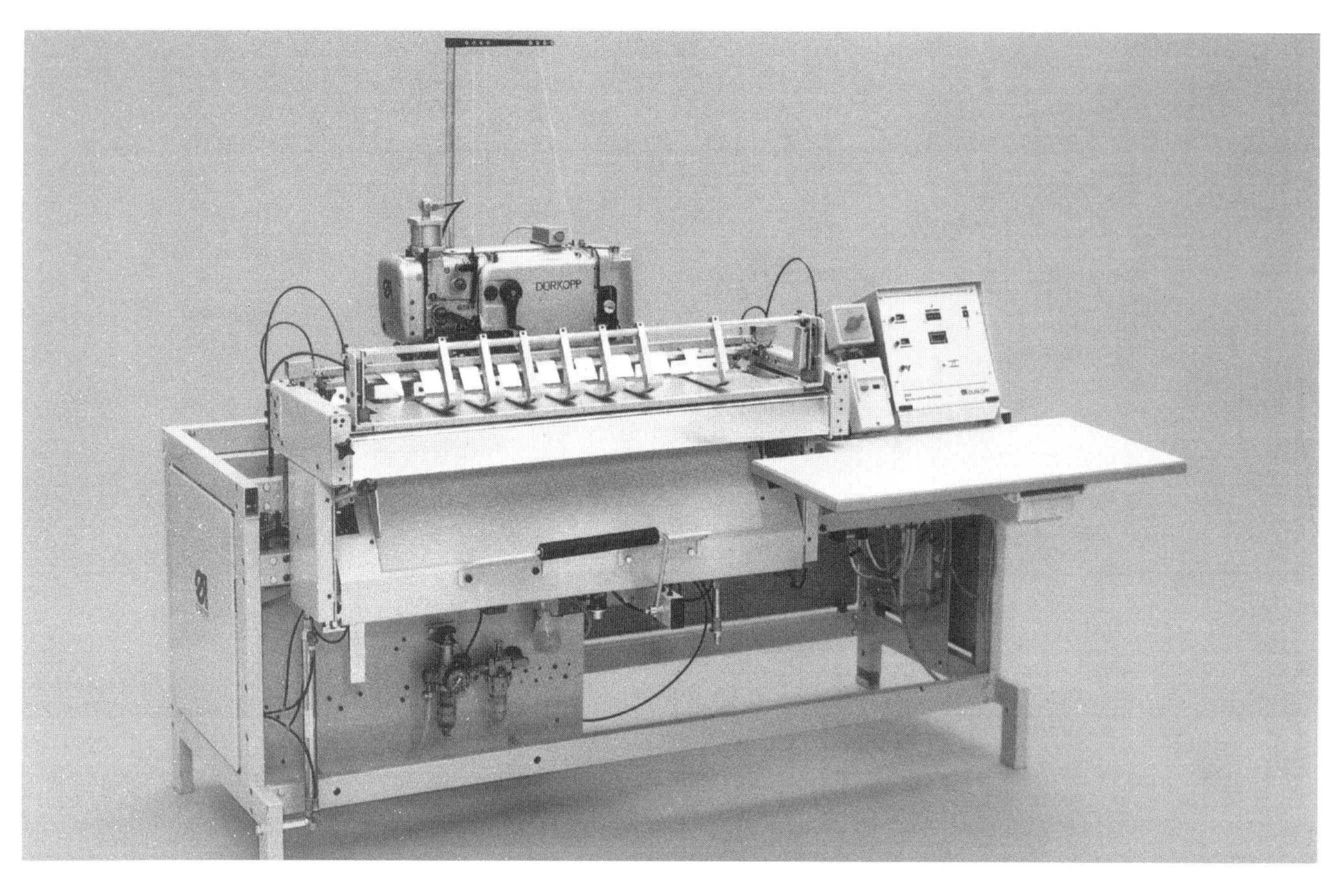


**FIGURE 38**
MICRO-ELECTRONIC CONTROLLED BUTTON-HOLE STITCHING MACHINE

The machine clamps the fabric and sequentially stitches the line of button holes on the front of shirts, blouses, etc., and stacks the finished pieces.

German factories they studied production was typically in small batches of between 150 and 300. Change-over between styles was frequent, and factories often produced a number of styles simultaneously. Despite the differences in output scale and variety, it was also found that average productivity was rather higher in the German factories than in the British. When comparisons were restricted to similar types of product, the number of garments produced per employee in the German plants was roughly twice that in the U.K. Overall, the *value* of the garments produced in the German factories was considerably higher – between 2.5 and 3 times higher in the case of export prices in two main product categories. The differences in volumes and in prices were related to differences in garment design:

> The spectrum of clothing manufacture covers an immense range – at the risk of caricaturing the gap between the two countries, it is worth describing in a few words the higher quality and styling of German clothing production. Three differences may be mentioned. First, the German product (and we refer here particularly to ladies dresses, jackets and suits) – in order to provide shape – consists of more separate pieces, and has more darts and tucks, to form a 'structured' and 'tailored' garment; secondly, it is more often made of checked or patterned material, requiring more skill in cutting and joining pieces together to ensure that the pattern aligns; thirdly, more decorative stitching and other detail (for example, pockets diagonally set in) is employed to provide interest and variation. The British garment on the other hand, is generally made of fewer constituent pieces, it is made of plainer materials, and it has less decorative stitching. Differences in quality of cloth and of trimmings were also apparent, reflecting the higher priced market for which German garments were destined.
>
> (Steedman & Wagner, 1989, p.42)

This comparison may appear to be largely a result of different manufacturing policies. What implications might it have for garment designers?

---

This is discussed in the following text.

Some important points follow from the comparison. Firms designing for long-run production of standardised products generally face direct competition from manufacturers in countries with low labour costs. Many German manufacturers have concluded that mass production of products with low levels of added value is a mis-use of high-cost, highly-skilled labour resources. So they have relocated this type of production to low labour cost countries. In garment manufacture, this primarily involves the relocation of the labour-intensive areas of assembly. Firms which have not taken this approach rely on two main factors – use of assembly automation where this is available, and maximising the advantage that is derived from proximity to the market. But automated machines tend to be inflexible in that they cannot easily be used for other types of production. Hence, the British factories in the Steedman and Wagner study were mostly dependent on a narrow design and product base. The successful use of automation in these circumstances is also dependent on very high levels of task specialisation, both in supporting the operation of automated equipment and in the other, non-automated areas of assembly. Operators are concerned with only one or two operations which are performed at high speed, and generally work on simple designs in order to maximise productivity. This de-skilling of the workforce is a further source of inflexibility in that operators cannot readily be switched to work on other types of task. Taken together, these factors present a considerable constraint on the degree of freedom in product development.

Marketing large volumes of low-cost, standardised products where unit profit margins are often low has long proved to be a successful strategy for U.K. retailers. But it is increasingly under challenge in intensive competitive conditions and as consumer preferences change towards higher quality, non-standard items. It is also questionable whether this has been the best long-term approach for manufacturers. Their competitive ability may have been impaired in a number of interdependent ways. One is a loss of design initiative to large retailers and loss of control of product portfolios. Related to this is a loss of flexibility which results from investment in dedicated equipment and limited workforce training. Investment in semi-automated production systems may confer some advantages in terms of production speeds and costs to manufacturers in the higher-wage countries. But this type of equipment is generally available in the world market, and its purchase thus gives buyers no long-term advantage since competitors can easily acquire *and* operate it, while it also constrains flexibility and freedom in developing new designs.

The possible precariousness of this strategy is indicated by the vulnerability of manufacturers to shifts in retailer policy towards 'sourcing' from low labour cost areas. The alternative approach, which was evident in the German factories looked at by Steedman and Wagner, is for manufacturers to identify sources of competitive advantage which cannot be readily replicated by competitors. Market proximity provides an element of competitive advantage, but to a diminishing extent as barriers to trade are reduced and as systems of transport and communications are improved. Instead, many German firms have placed reliance on strategies based on the integration of design, marketing and manufacturing. They have invested in all these areas to develop a reputation for design sophistication comparable to those of many French and Italian firms. The objective in this type of strategy is to develop high-level competences in design and manufacture which few competitors can match, and to target designs on market segments which offer high levels of **added value**.

To achieve this requires a comprehensive range of skills – in the various stages of design, but also in other areas. High levels of shopfloor craft skills are needed to ensure that advanced designs can actually be produced to the high standards required. Inevitably, this involves high levels of investment – but with an emphasis on the continuous development and updating of employee skills rather than on capital equipment. Without such in-depth and up-to-date skills, a company's overall freedom to explore new design possibilities is likely to be highly constrained. Unsurprisingly perhaps, the Steedman and Wagner study also found that the differences in performance between the British and German factories were paralleled by differences in the extent and standards of training. (It should be added that this study is one of a comprehensive series of comparative studies of different areas of manufacturing in Britain and Germany. The findings of these studies are broadly similar in terms of the extent of the German–U.K. skills gap and its consequences in terms of the types of product which can be manufactured.)

The strategy outlined above is not a unique, German phenomenon. The Japanese textile industry manifests a similar selectivity in that product areas offering low levels of added value per unit have been vacated, often by relocation of these activities to other parts of South-east Asia where labour costs are low. Production within Japan has also been concentrated on products which are 'knowledge intensive'. This is partly an outcome of high levels of expenditure on R&D which have resulted in product innovation. But it is also a function of in-depth levels of design capacities and of craft and other skills. In the clothing sector, where Tokyo has become one of the world's fashion centres, there has been a similar ability to differentiate between products which cannot be defended because of low skill content, and those which, at least, offer the potential for the development of **design-, knowledge- and skill-intensive products**. An even more striking example is provided by the Italian textile and clothing industries. These have built on their reputation for design excellence in some areas, and on the strengths of long-established geographical concentrations of particular crafts. The success of this is indicated by Italy's $10 billion surplus on the balance of trade in textile and clothing products in the mid-1980s.

## 4.5 THE SUPPLY CHAIN

Let us return now to the exercise in Section 3, and consider Question 9 as the starting point for this section.

All the materials in your selection (other than some ancillary items) are **polymers** – materials which are formed as long-chain molecules (Block 5). The main groups of textile polymers are shown in Figure 39. As can be seen, many are naturally occurring polymers, as in the case of cotton and wool and the cellulose-based manufactured fibres.

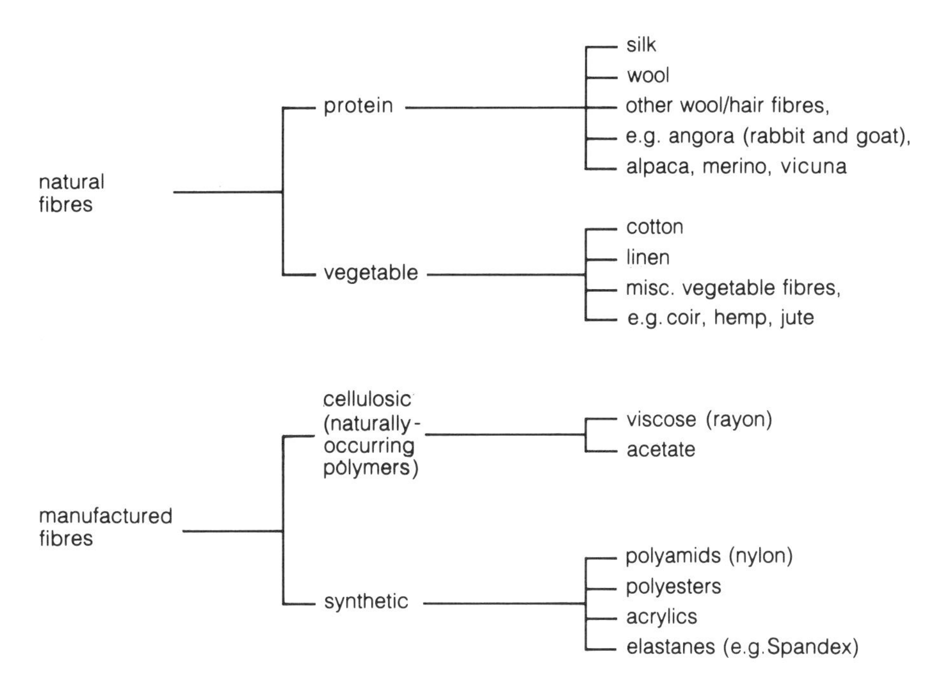


**FIGURE 39**
MAIN GROUPS OF TEXTILE FIBRES

In the latter group, textile fibres are produced from natural cellulose which is mainly processed from timber sources. The group of fibres classified as synthetics are mostly derived from petrochemical feedstocks via intermediate production of monomers – basic molecular groups from which polymers are subsequently produced.

Each of the main types of natural fibres – cotton and wool – constitutes a general category within which there are variants, distinguished in such aspects as fibre length, and linear density ('tex') which make them appropriate for different types of end-use. For instance, there are considerable variations in the wools produced by different types of sheep, between the same breed of sheep reared in different conditions, and in the wool from different parts of any individual sheep. Consequently, one of the first tasks in processing wool from any one source is that of grading – sorting a fleece into different qualities for different uses. Similarly, cotton varies with the variety of the genus *Gossypium* which is grown, the agricultural characteristics of the area of production and the methods of cultivation – differences which contribute to variations in fibre length, strength and tex. Figure 39 lists a number of other types of natural fibre – but it does not represent an exhaustive list of possibilities. For example, Rudofsky (1947) refers to garments made from fibres derived from banana plants. The natural fibres we now use are the product of a long process of selection, of improvement in plant and animal species, and in methods of cultivation or husbandry.

The manufactured fibres such as rayon, polyester and nylon are also more accurately regarded as categories of products, because they can be produced in a variety of configurations, for instance, as a continuous filament or as a staple of any pre-determined length. They are also produced in a range of cross-sections (see Figure 40), in a crimped form (for instance, mimicking the crimping of wool), and with additives such as colour molecules or titanium dioxide to modify the light reflectivity of some synthetics. The manufactured fibres are thus highly versatile, and capable, within their range limits, of precise specification for given intermediate and end-uses.

**FIGURE 40**

CONTRASTING PROFILES OF SYNTHETIC FIBRES

Profile design is an important determinant of the properties of fabrics and finished garments or other products. For instance, profile affects fabric weight and insulating properties, its handle, and wearer comfort, such as by reducing the area of direct contact with the skin and related friction levels. However, profiling can also involve penalties such as loss of extensibility and higher levels of dye use.

In some important respects, the properties of manufactured fibres tend to *supplement* rather than match those of the natural fibres. For instance, synthetics generally have greater strength, but they lack the hydrophilic property of the natural fibres. It is the latter property and the greater surface irregularities of natural fibres which influence contact with the human skin which are thought to be two of the more significant differences which contribute to the consumer preference for natural fibres, for some purposes (Gohl & Vilensky, 1983). (In essence, natural fibres remove skin moisture more efficiently, and their irregularities produce intermittent skin contact at a microscopic level, enhancing feelings of comfort.) These properties have been difficult to either replicate or otherwise imitate in manufactured fibres because of fundamental differences in structure at the level of the individual polymers, in terms of their bonding and the related structures which are formed.

Accordingly, while you will probably have found some products made from only a single fibre type in your identification of materials, you are also likely to have found that combinations of fibre types are frequently used. For instance, you may have found cotton–polyester mixes in shirts and sheets, and fibre mixes which include elastanes (highly extensible fibres) in stretch items such as socks and sportswear. Such mixes represent the selection in the design process of a combination of fibre properties – and costs. Variation in the properties of natural and manufactured fibres is one of the factors you may have thought about in relation to Question 9 (c) in the Section 3 Textiles Exercise. One of the other factors is availability. Fibres such as cashmere and vicuna are used in very few products because they are in short supply, and because they are very expensive.

The differences in properties of the various fibre types are reflected in considerable variations in the shares of manufactured fibres in the markets for specific categories of apparel. They account for more than 90% of fibre use for items such as tights and swimwear to below 30% for men's and women's trousers, workwear and undergarments (Davies, 1987). Reflecting back to Question 2 in Section 1.2, the increasing use of synthetic fibres in an expanding range of products stands out as the most significant change, compared with twenty and forty years ago. This has been evident in new fabric types, and in characteristics of strength, durability of shape and colour, stretch and lightness which were not available in the past. Fabric quality has increasingly matched that of high-quality natural fibres such as silk.

What might be the implications for designers of such changes in the types of materials that are available?

---

The capacity for modifying fibre properties and for otherwise innovating in fibre types lies at the heart of the longer-term cycle of product development and design in the textile chain as a whole, bearing out the point in Block 2 that 'suppliers of raw materials, component parts and manufacturing equipment are capable of changing a firm's range of products' (p.69). However, the flow of materials and components should not be regarded as a 'random factor' in a firm's product development programme – although it often appears to be treated in this way. At the least, designers need to actively undertake '*boundary scanning*' to maintain an up-to-date and comprehensive information base about developments in materials, processing methods and so on (Block 5). Equally fundamentally, firms may need to co-ordinate product development from the earliest, idea-generating stages with other firms upstream and/or downstream from their location in the production chain.

Recognition of the significance of this basic point, and the development of appropriate **strategies of co-ordination** between suppliers and buyers in a production chain, is another factor which has underpinned the performance of many Japanese companies in international markets – as you saw in the example of **'lean' production systems** in Block 5. In the textile supply chain, interaction between buyers and suppliers has long played an important part in product development and design in the past, and it is likely to become far more important in the future – increasingly on the co-ordinated pattern of the lean production model. Examination of some examples of past interdependences in product development reinforces some of the points concerning the innovation process which were considered in Block 3. These form the main concern of the rest of this section. In Section 4.6, we will look at the emerging pattern of development and how this is likely to affect the design and development of end-products. First, an outline of some of the basic characteristics of the chain.

The main stages in the **textile supply chain** are identified in Figure 41. While useful in its imagery of interconnecting and interdependent links, the notion of a 'chain' is also potentially misleading in that it may appear to suggest a single and simple flow down through the various sub-sectors or stages. In reality, each stage, from fibre production downwards, contains a range of product types which, ultimately, link to the diverse range of end-products. Generally speaking, the route through the chain for any given product depends on the particular end-use (product type, market segment and so on), and consequently on such variables as the type of fibre or fibres used in the main construction, the yarn type, and the method of fabric formation.

Thus, while the 'chain' analogy is frequently used to refer to the progressive transition from fibres through to end-use, this can be misleading in that there is no single, 'typical', flow down the chain. The progression can vary considerably, even for the same type of product. For instance, depending on the product, developments in dyeing technology have made it possible to add colour at most stages from synthetic fibre production through to final manufacture. Some garments can be made up from knitted or woven fabrics which are in a 'greige' or grey form (i.e. unfinished in terms of dyeing and in other respects), so

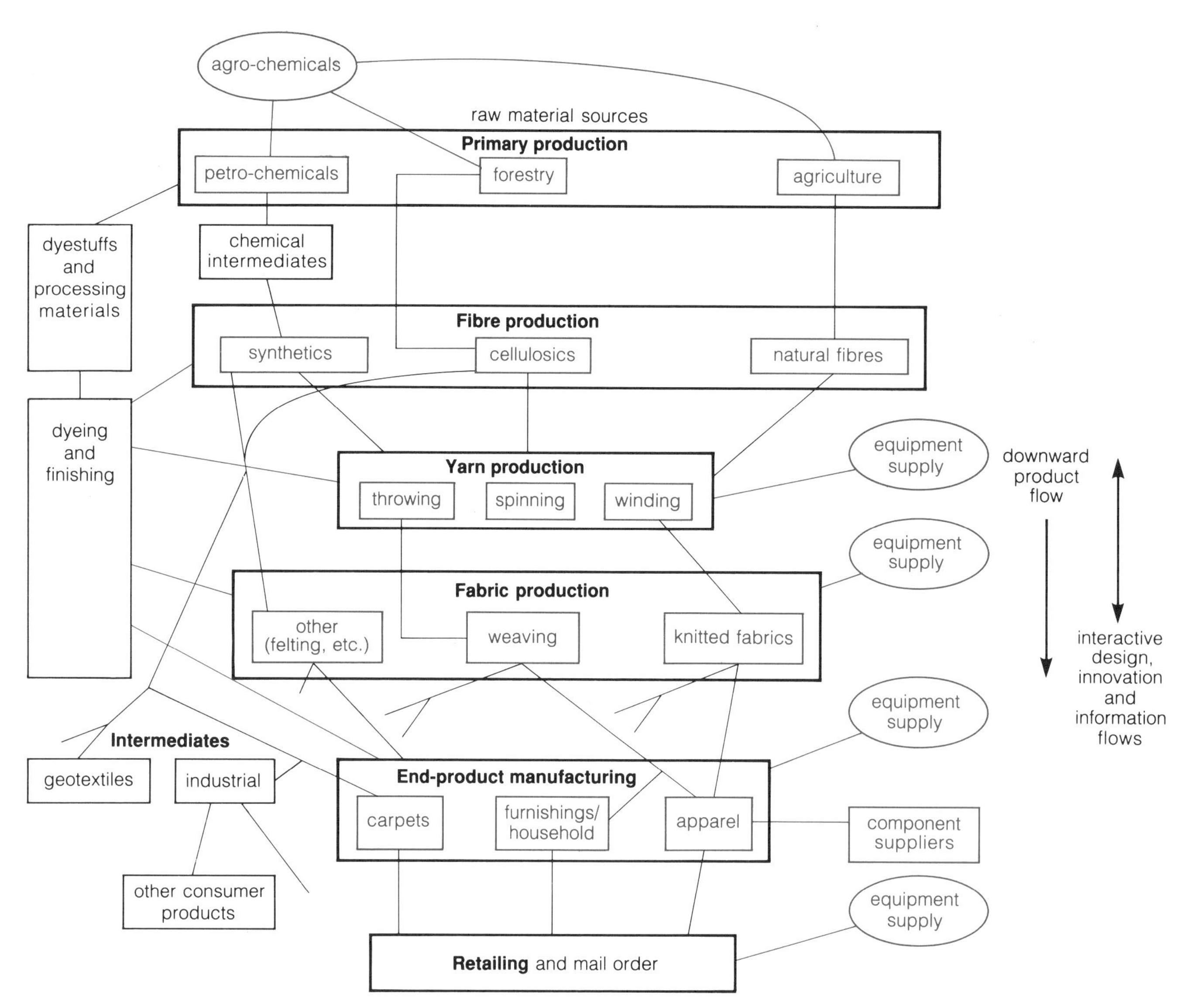


*Note:* Throwing and winding are two of the specialisms within yarn production.

**FIGURE 41**

MAIN COMPONENTS OF THE 'TEXTILE CHAIN'

**Notes on Figure 41:**

The main **flow lines** shown in the diagram are partly downward, as in the progressive processing of materials into retail products; they are also interactive, particularly in the information flows between companies, which relate to ordering, product development and design.

The **agrochemicals industry** is an important contributor to production in the textile chain, but less directly, i.e. through the flow of pesticides, herbicides and fertilisers to the primary producing industries for cellulosic and natural fibre production.

**Dyeing and finishing processes**, which are an important source of product innovation, are undertaken at any stage from fibre production onwards, but the pattern varies considerably.

**Component suppliers** are shown in the figure to indicate the role of a variety of specialist suppliers of such things as buttons, zips, trims, interlinings, etc.

that dyeing and related finishing processes constitute the final stage of production. An increasing number of manufacturers have turned to this method for products such as sweaters and basic (anorak style) jackets since it enables them to improve their responsiveness to market trends. Stocks can be held as greige garments and dyed shortly before despatch to retail outlets in response to the emerging pattern of consumers' colour preferences.

Two further basic characteristics of the textile chain need to be born in mind. First, while there are some very large, vertically integrated textile conglomerate companies (i.e. active in many stages of the chain), the various stages of processing have tended to remain separate in terms of their commercial functioning. Thus, the textile chain is also a large-scale trading system between buyers and suppliers. But, as will be discussed later, this trading system also provides information flows which shape the longer-term patterns of product innovation. The second characteristic is that, as in other sectors such as automobiles, the production chain has become international in its functioning. The supply and buying of materials and intermediate and finished products takes place on an international scale – as the labels in your textile products will probably have indicated.

## INNOVATION IN FIBRE TYPES

Since the late nineteenth century, the 'pre-industrial' natural fibres such as wool and cotton have had to compete with a growing range of manufactured fibres. The development of new generic types of manufactured fibres dates from the 1890s and the production of the cellulose-based – or rayon – fibres. The **primary generator** in this development was the desire to produce a manufactured silk – and rayon was for a long time referred to as 'artificial silk'. Silk was a target both because of its innate qualities and because of its scarcity and consequent high value.

Among the synthetic fibres, three developments stand out as major, **radical innovations**: polyamid fibres (i.e. nylon, in the 1930s); polyesters (in the 1940s); and elastanes (in the 1970s). In each of these cases, the initial development of a new fibre type has been followed by market success on a prolonged timescale, and has depended for commercial success on a variety of less dramatic, but cumulatively vital, **incremental innovations** which, for instance, have led to price reductions and to improved fibre properties. Although the industry is dominated by large science-oriented firms such as Du Pont, Celanese, Toray, Hoechst and ICI, all of which invest heavily in R&D, development from the early prototypes has been characterised by its length. An example of the timescales involved is provided by Figure 42. This charts the numbers of process patents registered for Caprolactum, one of the chemical intermediates used in polyamid production. As can be seen, the peak of patenting activity was reached more than thirty years after the initial patents for nylon production and for Caprolactum. (This figure also illustrates the conceptualisation of the industry life cycle in Blocks 2 and 3, where it was suggested that process innovations become increasingly important as a product matures.)

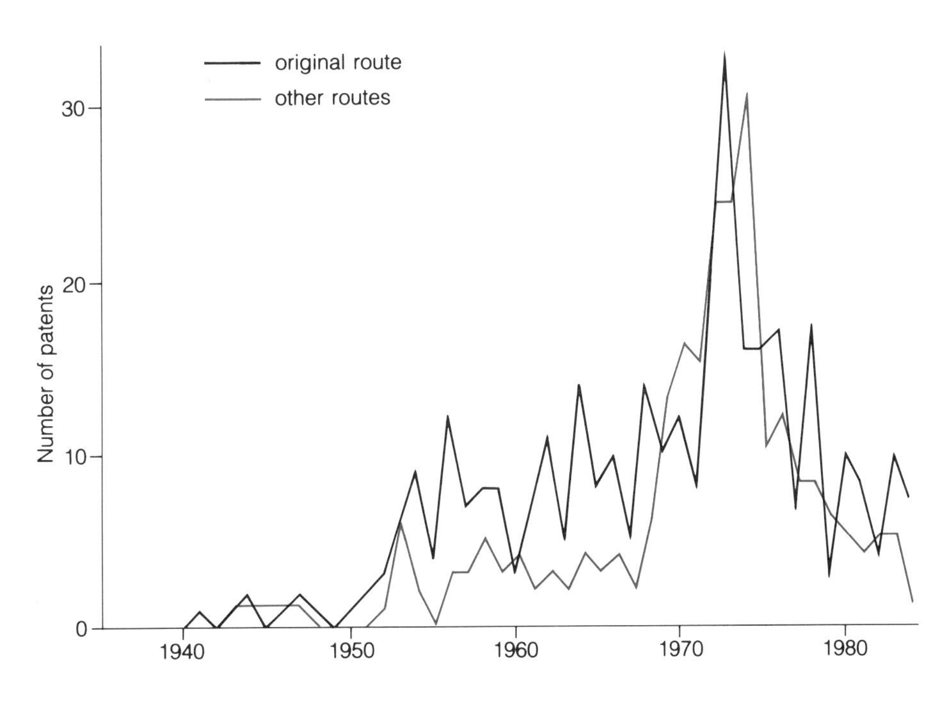


**FIGURE 42**
EXAMPLE OF THE TIMESCALE FOR INNOVATION

Graph of continued patenting activity for the production of Caprolactum, a chemical intermediate used in polyamid production

If you think back to the Block 3 study of bicycles, that stands in sharp contrast as a sector in which development has largely come from individuals. Allowing for the undoubted differences in the nature and scale of these two very different types of product evolution, it nonetheless seems that the Block 3 model of innovation corresponds closely with the pattern of development of manufactured fibres, and more generally with long-term trends in innovation across the textile chain. The following two examples illustrate this.

Polyamid fibres appear to have reached maturity, as is indicated by their declining market share – from 50% of world synthetic consumption in 1967 to 27% in 1986 (Anson & Simpson, 1988). Yet development of new product types has still continued, for instance, in the methods of texturing, in fibre profiles and so on which were referred to earlier. But an important factor in this is the dependence of many such developments on advances in other parts of the chain. For instance, the early use of synthetics in clothing and soft furnishing was constrained by difficulties in dyeing. Advances in dyestuffs were necessary to produce acceptable colour matches, colour fastness, and so on.

Similarly, a desired goal was to manufacture fibres at very low tex levels to compete with the ultra-fine natural fibres such as merino because of the high premiums these fibres are able to command for the production of ultra-fine, lightweight fabrics. Yet, while it has long been possible to produce microfibres, they could not be processed satisfactorily on the available spinning equipment. Advances in spinning systems enabled some reduction in the linear density of fibres that could be processed, but there were still limits which were imposed not only by spinning systems, but also by what it was possible to weave.

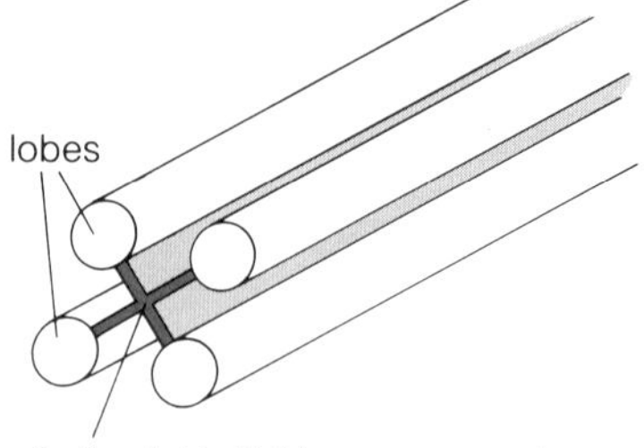


(A) fibre profile during spinning and weaving

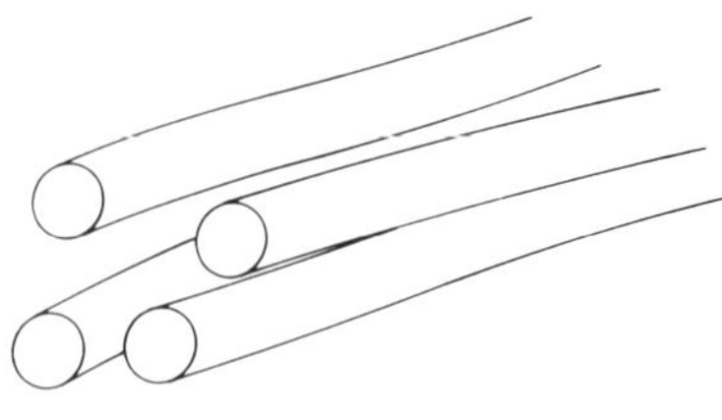

(B) after processing the 'lobes' become individual microfibres, following removal of the linking component

**FIGURE 43**

PRODUCTION OF MICROFIBRES

The fibre is initially manufactured in 'multilobal' form (A) consisting of the final fibre components and a linking component. The larger, dual-component fibre enables spinning and weaving to be undertaken – the established production processes in these stages cannot viably handle very fine fibres. The linking component is dissolved in alkali, and is removed after fabric production, leaving a finished fabric constructed from very fine fibres (B).

This **innovation gap** was eventually overcome by the development of 'split fibre' systems. As is illustrated in Figure 43, these involve the formation of fibres on a profile in which a number of microfibres are initially formed in groups, and are linked together by a component that can be dissolved in an alkali solution at a late stage in production. Thus, for spinning and weaving, the 'compound' fibre is used in a suitably high gauge, after which it is reduced to its sub-components of individual, very fine microfibres by treatment with a mild caustic solution. The resulting fabrics are suitable for the higher-value areas of the market, primarily for clothing, and have led to the development of new market niches. Such examples – of which there are many – emphasise the impetus from 'technology push' which underpins the longer-term processes of product evolution. More generally, developments in the fibre-to-yarn part of the chain are reaching the point at which it is possible to identify desired fabric properties and to then develop fibres, yarns, weaves and processing methods which will, together, meet this specification. For instance, some of the silk-like polyester fabrics available in the early 1990s resulted from this approach.

But incremental development of new products and processes has not been confined to the manufactured fibres. Competition has stimulated improvements in the cost and other characteristics of the main natural fibres, wool and cotton, and these improvements have limited the erosion of their markets. (The share of cotton in world textile consumption fell in the period from 1950 from more than 70% to 48% in the 1980s, but cotton output continued to increase because of rising world textile demand.) In part, the improved competitiveness of natural fibres is a result of changes in farming methods such as in sheep breeding and rearing, and in cotton cultivation and harvesting. There is also a capacity for switching production towards more profitable areas. For example, among other developments, the growth of the market for microfibres was followed by a shift towards raising merino sheep by the Australian wool industry.

While changes in agricultural technologies have primarily affected the cost of natural fibres and, to a lesser extent, their quality, there have also been improvements in their technical performance. The advances in theoretical knowledge which led to the development of manufactured fibres have provided new insights into the underlying chemistry of natural fibres. Among other things, this has contributed to improvements in fibre and other textile processing. Progressive developments in dyestuff types, in resins and other chemical products, have made possible an increasingly precise modification of textiles which, in some cases, extends down to the most fundamental level, that of the microstructures of individual fibres. For instance, these developments have allowed more consistent dyeing – i.e. *replicability* of precise colour standards – and the modification of textile performance in other respects, such as improving shape retention in garments through enhanced crease resistance, texture and shrink resistance – the latter being particularly important in the case of wool products. Behind this is the more general point that the development of new materials or product types often acts as a stimulus for the improvement of established materials or products which are challenged by the newcomer.

**SAQ 11**

Suggest three examples of products (not necessarily in the field of textiles) where design improvements appear to have been stimulated by competition from radically new types of product.

## 4.6 INTEGRATION: A PATTERN FOR DEVELOPMENT?

As you saw in Section 2, the increasing share of products marketed by the large retailers, particularly as 'own-label', has passed much of the initiative in product development and in garment design from manufacturers to the retailers who, to varying extents, employ their own design teams. In the branded goods sector of the market, manufacturers retain the design initiative, but where they produce own-label products for retailers they effectively become sub-contractors. Design, particularly in the earlier stages (conceptual and embodiment) then becomes an interactive process of (more or less) **creative tension** between designers and others making design decisions in the retail organisations, and the designers or design teams of manufacturers who are bidding for contracts. Initial design ideas and concepts may emerge from retailer and manufacturer, be further developed by the manufacturer, and are then considered for inclusion in the retailer's range. What this does seem to ensure is that there is broad exploration of the design possibilities. This is indicated by what appears to be a fairly usual experience among manufacturers – that somewhere between 10% and 20% of designs offered to retailers are accepted, and then, often only after considerable discussion and modification.

In effect, what emerged from these developments, in both Britain and the U.S.A., were **retailer-driven production systems** which dominated the mass sectors of the market. In these systems, key design decisions and many of those relating to production – volume, cost parameters, timing and so on – are taken by the retailers rather than the manufacturer. But this has produced some generally acknowledged problems. The low success rate of manufacturers' design submissions carries considerable costs since each garment concept which the manufacturer and/or the retailer wishes to push forward as a production possibility has to be made up as a complete sample garment so that it can be evaluated. In the long term at least, this inflates production costs, while this process also contributes to long development lead times.

However, this problem is moving towards resolution. Computer-aided drawing systems provided an important starting point in changing approaches to development in the concept stage (Figure 44). Among other things, these systems permit the rapid exploration of a wider range of design possibilities than is feasible using traditional methods. Subsequent systems have allowed highly accurate simulations of the style and finish of the manufactured fabric, and the modelling of their application to specific furniture and clothing styles.

What might the implications be for the role of the designer and the nature of design tasks? (Think about this for a few minutes.)

---

Application of current generations of CAD is enabling the design of fabric and garments to move from separate functions towards integration. A single design team can generate fabric designs and garment designs simultaneously. This locates design at the hub of the overall system of product development and manufacture – the information generated in such an integrated design group provides inputs for direct upstream use in fabric production, dyeing and so on. Equally, elements of the information flow can be used directly downstream in the preparation of, say, catalogues and promotional brochures. In the case of clothing design, the next step is towards dynamic 3D modelling. For instance, one of the concerns in the design of many women's garments is with the way that combinations of styling, fabric and cutting will 'hang' on the wearer, and how they behave in movement. As you saw in the consideration of dynamic modelling in Block 4, developments in CAD permit complex simulations. This potential is increasingly close to being added to the inventory of tools at the disposal of clothing designers. The parameters involved in garment simulation have presented some particularly difficult problems linked to the issues of measurement, fabric behaviour and so on, which were discussed in the previous section.

These developments have made it possible for a large element of design evaluation and modification to be undertaken prior to the final stages of decision-making in which made-up samples need to be used. In one case, the retailer's objective in moving towards initial evaluations based on designs on-screen and on high-quality print-outs has been to lift the figure of acceptance of designs at the sample garment stage from around 10% to 80%. The next step in development is to achieve links with the subsequent design tasks in which, as you saw in Section 4.3, CAD systems are already used comprehensively, and to strengthen links with manufacturing. As an example of this, it is now possible to link CAD systems to laser-printers and to ink-jet printers for the rapid production of either high-resolution printed reproductions of garments or a fabric sample from which a prototype garment can be made. In either case, the time involved in the process of design and selection has been greatly reduced, potentially eliminating some two to three months from the total development time.

**FIGURE 44**

CAD SYTEMS USED IN THE CUSTOMER INTERFACE

Apart from their role in designing, CAD systems also allow the rapid display and comparison of garment designs with different print or colour combinations. Among other things, this can contribute to reduced time for customer decisions.

However, there are dangers that the potential of applications like these will be viewed in a restricted way – it is not only designers who can productively utilise techniques for creative thinking! More generally in U.K. industry, many company managers have viewed the introduction of CAD largely in terms of a potential for reducing design costs. In particular, managers have been concerned to reduce the numbers of designers and design labour costs, perceiving the process in terms of 'design automation' (Simmonds & Senker, 1989). Accordingly, CAD installations have been justified financially on the basis of expected savings in design labour costs. In practice, these savings have rarely materialised – and this failure may have acted as a deterrent to other CAD investment. But this results from a narrow view of the potential gains from CAD. The real benefits have come from greater freedom and speed in the exploration of design possibilities. Other benefits have followed from the capacity for data transfer for use in production, in production planning, and in cost control. Overall, these benefits have enabled shorter lead times, more rapid changes in product lines, and increased diversity in product ranges, all of which allow increased emphasis on niche marketing. However, as was indicated by the Anglo-German comparative study referred to earlier, achievement of these design-based benefits is partly dependent on changes in manufacturing skills and methods.

In the clothing industry, utilisation of CAD has had a labour-saving impact in the areas of pattern grading and lay planning (Section 4.3). But, taking CAD applications as a whole, it has only gradually been appreciated that they also provide a means towards fundamental changes in business strategies and methods. In the context of highly competitive conditions and static demand which was considered in Section 2, retailers and manufacturers in the U.K. and elsewhere are following the trend in other industries, and moving towards 'demand-driven' systems.

Rather than manufacturing a large part of a production run before the product goes on sale, the objective is to produce only a limited proportion of the predicted production volume. Ideally, only 10 to 20% of the expected volume should be produced before an item goes on sale. In such **'quick response' systems**, subsequent production orders are triggered by the volume of sales at retail outlets. Where successful, this inversion of the traditional approach to production has two main types of aim. One is cost-oriented. In part, this should follow from an overall reduction in the levels of stocks of particular items in the shops and in retail distribution centres. There should also be a reduction in losses which result from faulty demand forecasting. For instance, 'stock-outs' where consumer purchases are lost because a particular size or style has sold out should be reduced, and the volume of products which have to be cleared in 'end-of-season' sales should be lower. The second concern is to improve levels of customer service by offering a wider range of products and of size and styling variants within any particular range.

However, to move from the 'traditional' pattern of production-led supply to demand-led systems will require changes at all stages in the production chain, particularly for large batch orders. The potential cost savings will not be achieved if stock levels are merely pushed to upstream levels of a production chain. As an ideal model, the whole production chain for any product should move towards operation with:

- low levels of stocks and work-in-progress;
- total reliability in meeting production schedules;
- flexible production systems, which give the ability to switch rapidly between products of different types;
- viable production of small batches of any product;
- short lead times in the planning and production of new product types;
- very high levels of conformance to quality standards.

As you will recognise from Block 5, this has close affinities with the *'lean production'* model. The objectives are similar at strategic and operational levels, although some of the problems, such as those related to timescales of product availability, are rather different. But the problems which have to be overcome in achieving the change-over are formidable. They include:

- high levels of co-ordination between companies operating at different stages in the production cycle (for instance, to be able to guarantee supplies at the required notice period);
- (in some cases) a need for more flexible types of production equipment;
- higher levels of skill among all groups of employees;
- an ability to discard long-held beliefs about 'best practice' methods, and to adopt new methods which may sometimes appear to be 'wrong' (for instance, managers accustomed to operating with the security of high levels of buffer stocks often find it difficult to operate without this security, and have difficulty in recognising its positive cost implications);
- achieving compatibility in hardware and software between different production stages and different companies, and a related willingness to accept the interchange of data for production purposes, for production control, billing, and so on.

In total, this amounts to some very far-reaching changes in approaches to product design and in other areas. If these changes are successfully achieved, domestic producers are likely to reap a new set of benefits from their proximity to the market. But, as earlier consideration of the nature of the general pattern of technological innovation has emphasised, a transition of this magnitude is likely to require many small advances towards the ideal model, and to take a very long period of time, possibly more than a decade.

What might be the implications for product design? (Think about this for a few minutes.)

---

Change along these lines will place greater emphasis on product design in both incremental adaptation of designs and in more radical terms. This follows from the associated broadening of product ranges, and from the greater competitive emphasis on design innovation which is associated with flexible production. In addition, the continued change-over to electronic design systems results in the outputs from the various design stages becoming a driving element in the wider, electronically-based integration of production. This integration extends back from electronic point-of-sale (EPOS) systems through all stages from raw materials production. One possibility is that the concerns of designers in different areas of the production chain will increasingly become integrated into a *total design activity*. This can include a broadening of the design area which does not only represent, but can extend to the design of, the 'functional environment' (Block 4) within retail outlets:

> ... garment designers are becoming more involved in fabric and component design. In selling a design concept, they have also made a contribution to the marketing aspects, how a garment might be packaged, how product information is presented, and how garments might look in a store environment. They are becoming a focus of the total design activity.
>
> (Walter, 1992, p.86)

This concludes the textile study. Apart from its links to earlier parts of the course, it has also identified a number of questions about the broader organisational and strategic contexts which influence product design. These issues are taken up within a more general context in Section 5.

**FIGURE 45**

SPEED TO THE NEXT SECTION!

The intrepid cyclist's assistant is mounting rockets as a speed booster. The combination of well established traditional clothing with innovation in bicycle design enabled the German engineer Richter to reach a speed of 90 km/h. He avoided serious injury when the machine left the track and exploded.

# 5 DESIGN AND ORGANISATION

As a conclusion for this Block, and for the course as a whole, this section, together with Section 6, is concerned with aspects of change in design practice which seem likely to continue to be important over the next few years. This section is concerned with changes in the methods and functioning of the organisations in which – or for which – product design is undertaken. This is related to changes in design practice which, on a wider scale, are comparable in magnitude to those you considered in Section 4.6. There are potential tensions between the pattern of organisationally-driven developments and the 'traditional' concerns and methods of designers. As will be seen, approaches to the organisation of design and of the related functional areas can be fundamental in fostering 'good' designs – by whatever measure – and in inhibiting them.

## 5.1 ORGANISATIONAL INFLUENCES

Earlier parts of this Block, and the course as a whole, have referred to some of the technological and organisational developments which are changing approaches to product design.

### SAQ 12

Think of three examples of these changes from your earlier work on the course.

Changes in technology and in methods of organisation are inevitably modifying the way in which design is approached and how designers work with people in other areas of organisations. Such changes are not always welcomed. One of the problems which designers have to confront is that of seemingly uninformed or unnecessary inputs or obstructions which emanate from other parts of a manufacturing organisation or, in a consultancy, from clients with uncertain or conflicting ideas. Undoubtedly, people outside the areas directly concerned with design are sometimes uncertain, vacillating, 'wrong-headed' or uninformed, but there is generally more substance to the tensions that often exist between designers and the various other groups who are concerned with the genesis and outcome of a project. Whether they like it or not, designers have to acknowledge that there are no clear boundaries between those formally concerned with product design and the many other departments and people who, in various ways, have an interest in design outcomes, and contribute to the shaping of those outcomes.

The ability to comprehend the patterns which underlie the links between design and 'non-design' groups, to recognise the nature of **organisational constraints**, and to work within (and around) these constraints is an important component in the designer's range of skills. Just as the availability of materials within a specific combination of price and technical characteristics can shape product design, so a range of factors within a manufacturing organisation can limit the parameters of design – or push them in a particular direction. For instance, past financial or other commitments to a particular type of material or manufacturing process (for example, steel car bodies, considered in Block 5) may limit the designer's freedom. Just as the capacities of materials need to be understood, and sometimes have to be 'stretched' in some way to make a design modification possible, so the complexities of the organisational system need to be recognised, not least because they may have to be carefully navigated in order to make a new design viable or possible.

**SAQ 13**
Organisational factors of various types have been referred to in a number of the earlier Blocks. One of these was Block 3. In what ways was it suggested there that organisational factors might influence creative design ?

**EXERCISE ORGANISATIONAL INFLUENCES**
Think back over earlier parts of the course and about your experience – whether working in organisations or in other types of context – to write down a few ideas about the possible influence on product design of the organisational context in which design takes place.

**Spend about five minutes attempting this exercise before reading on.**

A response to this activity will be developed in the sections below. However, as an initial framework, a start in envisaging the impact of the wider organisation on product design is provided by the 32 'elements of a complete product design specification' set out in Block 2, Figure 45 (p.105). The impact of many of these elements – such as 'manufacturing facility', 'legal' and 'market constraints' – is likely to come via other departments in a company, and thus to reflect the particular views and concerns of those working in these departments. The impact of the wider organisation needs to be considered in three main respects:

- those which may more directly support or inhibit the creative and other processes involved in design;
- that of the multiplicity of specific concerns and interests related to a particular design project – such as those related to component supply, the financial viability of the product development, and aspects of manufacturability;
- the broader aspects of company policy which drive product development and design in particular directions – most obviously, the market and other strategies of a company (see 'Strategic options', Block 2, pp.72–73).

These points are developed further below.

## 5.2 THE ORGANISATIONAL SETTING

The need to adapt forms of organisation as a part of technological change is not a new phenomenon. Block 1 drew attention to some of the disjunctures that followed from the first Industrial Revolution in the eighteenth century. This 'revolution' is generally thought of (and taught about) in terms of the invention and introduction of new types of machinery which enabled products to be made more cheaply and in greater numbers. But the successful introduction and utilisation of this machinery was generally (though not invariably) inter-dependent with innovations in forms of organisation. Most frequently, this took the form of concentrating production in factory units in place of a variety of home-working and sub-contracting systems. (However, this is emphatically *not* to suggest that given types of technological hardware are necessarily associated with particular methods of organisation!)

The shift towards factory production required more conscious attention to product design than had been the case before. The essential, underlying point is that factory production presented a different set of design problems. It was found that, in moving from the traditional methods involving **custom-building** individual items (see Block 1) and

**small batches**, designs for products which were to be made in **large batches** had to be adapted to the constraints of the manufacturing system and, in some cases, to the different markets that were opened up by higher volume production. For instance, while it is feasible to machine parts to conform to complex component designs when producing a **'one-off' product** or a **prototype**, this may be prohibitive on cost grounds for **volume production** (see Block 5), or it may not be possible to produce the part with highly specialised production equipment, and limited availability of shopfloor skills. But it is not only the nature of design problems that has changed. Design has also become a more formally directed and organised process as manufacturing has grown in scale and complexity.

What has emerged today as the characteristic organisational context for much consumer product design is the large – often the *very* large – company, operating many factories, and competing at an international level with other comparably large companies. In both the U.K. and the U.S.A., over one-third of manufacturing production is accounted for by the largest 100 firms (Dicken, 1986). In large firms like these, individual projects for product development tend to be located within a wider 'portfolio' of projects, each of which has to compete for financial and other resources. In many cases, a considerable number of other companies – large, as well as small – are tied into the design, production and other processes of major manufacturers, for instance, as suppliers of materials, services and components. Factors such as the scale of resources needed for product development and the intensity of competitive pressures have meant that, to an increasing extent, only large, established firms have the capacities needed to successfully manufacture and market new products on a national and international scale – and even they face difficulties, as will be seen below.

## MARKET LINKAGES

However, it is not only the availability of resources for development and the assets for their production that are important. As you have seen in Blocks 2 and 3, the success of new products depends to a considerable extent on the effectiveness of their 'coupling' to the 'needs' of the market at which they are aimed. Where new designs involve the replacement or modification of successful, existing products and thus slot into established patterns of distribution, this may be unproblematic. But when a radically new design is developed, or an existing design is aimed at a new market, difficulties may be encountered, even where design and manufacture have apparently been highly successful.

### SAQ 14

Give examples from earlier parts of the course of innovative designs which failed, or had difficulty in finding a market. How do these appear to differ from examples of innovative products which succeeded?

The difficulties can arise from misjudgement of market demand and/or from inadequate linkages to the new market. Teece (1986) explains these linkages in terms of control over, or access to, what he terms **complementary assets** and **capabilities**. In addition to production systems, these assets include marketing, distribution, sales and after-sales support systems. Any of these assets may be highly specialised, and their development may be both lengthy and expensive. For instance, Teece provides the examples of the introduction of a new drug requiring the dissemination of information over a specialised information channel, and the need to develop specialised repair facilities before the rotary car engine could be successfully marketed. As you will see in Section 6, the successful introduction of electric cars would similarly require the development of appropriate 'complementary assets'.

Lack of access to appropriate complementary assets, or ventures outside the sectors in which complementary assets are held, can be a source of market failure. Think back to the example of the first Moulton bicycle in Block 3. While Alex Moulton succeeded in manufacturing his bicycle in the face of resistance from the established firms, it ultimately fell victim to (seemingly) inferior competing products which were backed by large-scale production and marketing. Teece provides a comparable example in the development of the body scanner by EMI. While a successful product in a technical sense, it lay outside EMI's normal areas of activity. Market success went to American companies which developed their own scanners (infringing EMI's patent rights in the process), and which were able to market them through their established sales network in the medical sector. However, the power and resources of large corporations in this respect do not provide certain guarantees of success – as was emphasised by falling market shares and financial losses by companies such as IBM and General Motors in the early 1990s.

### DESIGN AND 'THE ORGANISATION'

So, as can be seen from earlier parts of the course, successful designs (at least, in terms of the product market) are dependent on many other factors within a manufacturing or other organisation. The importance of 'complementary assets' provides only one example of such interdependence. But how does design fit within the overall pattern of organisational functioning?

The answer, of course, varies from one organisation to another. There are no universally applicable solutions to the many problems of organising and co-ordinating the activities of large numbers of people in complex, multi-activity organisations, and approaches to organisation vary considerably. For example, some companies are highly centralised while others favour decentralisation. Similarly, some companies operate a 'command and control' approach to organisational authority while others rely on consensual approaches. But there are some commonalities. Most companies are hierarchical – with carefully graded levels of authority and status. All have some system of division of responsibilities which spans all areas of activity – from financial control through to the organisation of production and other functions, and to the administration, supervision and control of employees. Hence, the activities associated with design, while vital for the survival of manufacturing concerns, amount to only a small part in the total system of organisation. Even within the process of product development or creation, design is located within a number of other major organisational functions and concerns – as was illustrated in Figure 49 in Block 2 (p.115). This means, of course, that the objectives and concerns of the design process are shaped by – and shape – a number of other organisational processes and concerns.

### SAQ 15

From earlier parts of the course, what other processes and concerns that interact with design can you identify?

Not surprisingly, a great deal of time and effort has gone into trying to understand differences in organisational structures and methods, and to discover which may be the most effective in contributing to successful long-term company performance in financial and other respects, to adaptiveness in changing market and other conditions, and so on. There are no simple and generally applicable solutions to such problems. A considerable range of factors may, either singly or in combination, constrain or nurture creative activities such as those involved in product design.

In seeking to understand the potential barriers – and thence, to surmount them – and in identifying factors which are conducive to successful product innovation on a large or on a small scale, one of the more fundamental points to recognise is that organisations are not simply 'structures' – of units, departments, and so on. Nor, often, despite the wishes and best efforts of those in overall control, do they function with the smooth unity of, say, a mechanical or military machine. The concerns and responsibilities of the various functional and other units within an organisation are not homogeneous. Obviously, the priorities and pre-occupations of, say, managers in a production unit will be rather different from those of, say, design engineers or marketing personnel. The people within a particular unit generally add further to these differences through their particular background of experience and of professional and occupationally-related ways of viewing and analysing issues and problems.

All of these factors contribute to varied priorities and perceptions of the many issues and problems which are faced within an organisation, and many of these will, at times, focus on issues surrounding the development of a new product. The way that differences in *priorities*, *interests* and *perceptions* tend to be resolved is through a combination of *discussion*, *bargaining* and *compromise* which, in effect, amounts to an internal 'political process'. In the context of new product development, this emphasises that it is to be expected that differing views about many aspects of a project will be encountered at *all* stages, and that what emerges is necessarily a compromise. However, the existence of divergent views and priorities does not lead inevitably to compromise. In some cases, groups which feel threatened by an idea or development – or which regard it as 'unsound' in some way – may seek to block progress. (Remember the example of the Advanced Passenger Train mentioned in Section 8 of Block 3.) This resistance will not always be open, particularly if channels for the expressions of interests and views are inadequate.

**SAQ 16**

The number of different interests involved in a project can be gauged from earlier parts of the course. In aggregate, these have identified a number of prerequisites for a proposed new or adapted product design to be accepted and manufactured, and for it to succeed. Identify four of the more important prerequisites. What do these suggest in terms of inputs to product design?

If you think back to the consideration of creative approaches to design in Block 3, this links to points in the discussion above. What might these links be?

---

You will remember from Block 3 Section 7 that it was suggested that 'discussion and teamwork' can be important in fostering ideas and in defining and solving problems. These can also be important in generating a 'synthesis of ideas' (Section 3.2), and in identifying the broad range of 'dissatisfactions' with existing products. Obviously, including the collective ideas and endeavours of, say, marketing, sales and production people in the early (and other) design stages is more likely to lead to a thorough re-appraisal and to a high standard of improvement in the new version, and they are likely to contribute in other ways to the genesis of creative ideas (for example, in divergent and associative thinking). Also, they will add to the span of knowledge and skills. However, bear in mind that it was also suggested that teamwork needed to be carefully approached to ensure positive outcomes.

An important ability that designers need to develop is to come to terms with this process of discussion and compromise, and to operate within it to achieve satisfactory results rather than, as is sometimes the case, to see it as being a barrier preventing 'good design'. Yet, compromises are often regarded as somehow inherently unsatisfactory and as falling short of what would be the 'ideal' solution (as it may appear to a single person or specific group). In contrast, it is often argued that the process of achieving a consensus about the most effective outcome is regarded as likely to contribute to the strength of the final result. While necessarily involving a compromise, since some positions will have been modified, the outcome will be the product of pooled experience, knowledge and information, and it is likely to be one that has a broad span of commitment and extends across the range of people and groups that may be involved.

But, whether in consensual or in other approaches to organisational control, it is still necessary to ensure that organisational policies, tasks and objectives of different types are accomplished to a (more or less) satisfactory standard. This is generally accomplished by a combination of direction by the hierarchy of authority and control, and by a sometimes labyrinthine range of rules, operating and other procedures, prescribed methods of practice, codes, and so on. Thus, up to a point, the various tasks involved in product design are, in larger companies at least, directed by a variety of procedures and rules.

Yet the various formal – laid down – procedures and rules are not the only factor. Beside them are the realities of what actually happens: the various short cuts, understandings, improvisations and deals which ensure that things do actually get done.

Over time, the people working within organisations with their accumulated, shared experience – together with a number of external factors, such as government policies, legal requirements, competitive conditions and so on – contribute to the development of 'organisational cultures'. In simplified terms, these consist of sets of shared values and assumptions which extend across the broad bands of technological and other activity. In part, these can be seen as **routines** – 'a set of ways of doing things and ways of determining what to do' (Nelson & Winter,

1982). However, not all of these routines extend across the whole organisation. Some elements are specific to particular occupational groups – such as designers, production engineers, and so on.

'Routines' in this sense can have a powerful effect in fostering the processes of idea generation and other aspects of problem exploration and the identification and evaluation of potential solutions, for instance, where they include an in-built propensity to search for new ideas and explore problems. But sometimes, routines can impose blinkers. For instance, remember the 'not invented here syndrome' from Block 3 – an example that appears to be fairly widely encountered in manufacturing enterprises in Britain and the U.S.A. Similarly, the routines developed by many companies for the financial appraisal of new product and other projects have been widely criticised as 'risk averse' and as excessively concerned with short-term results and the neglect of longer-term considerations of product and business development. By contrast, the 'routines' developed in some successful Japanese companies appear, among other things, to have favoured more open approaches to new product generation. Specifically, there is a readiness to borrow the ideas of others – but then to improve on them.

'Routines' are important in another respect. Together with wider sets of commitments – such as those of past investment in equipment and in the development of sets of knowledge, skills and expertise, they serve to focus the search process in product development and in other respects. In effect, what tends to happen is that the process of search – the exploration of problems and the investigation of potential solutions which surrounds much of the creative process in design – tends to be directed along what have been termed **avenues** or **trajectories** – see Figure 46. The direction of the search process within parameters that are largely set by past commitments (rather than by conscious decision) is an important factor in explaining the predominance of incremental innovation. While the avenue or trajectory provides a base of established achievement and practice on which new product variants can be built, it can also present a constraint on thinking.

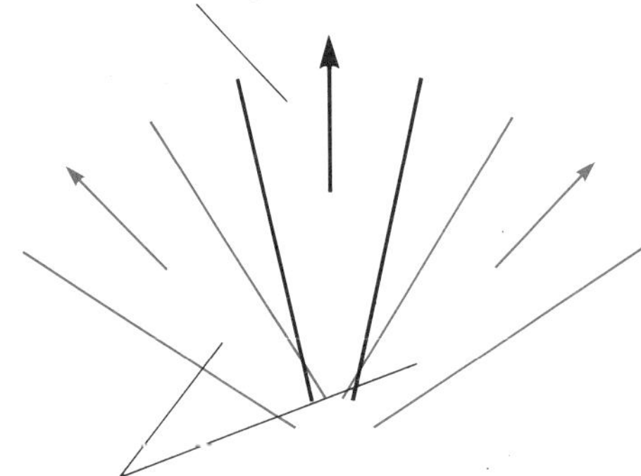


**FIGURE 46**
SCHEMATIC REPRESENTATION OF EXISTING AND ALTERNATIVE DESIGN TRAJECTORIES

In particular, readiness to consider possibilities which lie outside the established trajectory may be limited. For example, think back to the contrast between Raleigh and Moulton. Similarly, when one of the many deceased U.K. car manufacturers was offered the designs, dies and tooling for the Volkswagen Beetle in 1945 as a part of post-war reparations, this was rejected as an unconventional design (rear-engined etc.) which, it was believed, would never succeed. Thus, an important ability which needs to be fostered is that of readiness to recognise the indications that an existing pattern of design development may offer diminishing rewards, and to identify appropriate new technological and other possibilities.

## 5.3 THE SHIFT TOWARDS FLEXIBILITY

If you think back through the rest of the course, in one way and another you have considered aspects of the design of products across a wide range of complexity – from chairs through to houses and cars. At a number of points, the evolution of design and of manufacture has been set in a historical context, both generally, and in examples of specific products – such as vacuum cleaners (Block 2), and bicycles (Block 3). At a general level, the range of historical and contemporary products which you have considered shows that the nature and context of product design have changed, and continue to change, in some very profound ways. From the antecedents of the pre-industrial era, there has been a shift from comparatively simple products, most of which were made as single items or in small numbers. As in the example of the farm wagon (Blocks 1 and 4), these were often produced from designs that were not written down and which had evolved slowly over long periods of time. Their creation depended on long-established craft skills and knowledge.

Nowadays, product design usually involves elaborate goods, as is evident in the complex 'engineered' materials which are increasingly used, in the sophistication of many product components, and in the methods by which they are manufactured and distributed to the point of sale. Most consumer products are made in large, sometimes very large, numbers. The development of new designs and subsequent manufacture often depends upon the combination of **formalised** and **implicit knowledge** from a variety of technological domains (most obviously, in the example of cars in Block 5). This knowledge is embodied in the product and in the various processes and stages of the production systems, the complexity of which has increased greatly. Furthermore, as you saw in Block 2, greater complexity is not confined to the design–production process, but has extended outwards in elaborate processes of market research and consumer testing as firms have sought to reduce the guesswork and financial risks involved in new product development. Correspondingly, the overall process of product development has increased in complexity and rigour, and has tended to become highly formalised.

The shift to mass production methods has roots in the nineteenth century, but it was most obviously manifested in the automobile production systems developed by Henry Ford (see Video 5, Section 4), although it became increasingly significant across a wide range of consumer product industries. However, while the period from the 1920s through to the present time tends to be thought of as an era of mass production, it is important to emphasise that a considerable range of manufactured products remained outside the standardised, mass production category. As in the examples of clothing earlier in this Block, craft-based systems remained important, particularly in specialised market segments. Piore and Sabel (1984) estimate that only about 30% of U.S. manufacturing production (and 25% of Japanese) comes from mass production systems. However, these percentages would be rather higher for the consumer product sectors alone.

Nonetheless, in combination with the growth of large, transnational companies, the dominant trend by the 1960s appeared to be towards a **global system of mass production**. Development appeared to be towards world products – designed for the total **global market**, and produced in 'world factories', many of which would be concerned with only a small part of manufacture, such as the production of a limited range of components. This trend was evident in a number of industries – from clothing to electronics and automobiles. In one major and highly influential research programme on the car industry undertaken during the 1980s, the initial belief of the participants was:

> ... that intense pressure for energy conservation and environmental protection would make the small or light car the standard-size vehicle in all the world's auto markets. This would allow this type of car to become increasingly commoditised; all cars would be small and all would look pretty much alike, yielding the 'world car'. A second assumption was that market place competition [...] would be increasingly based on price and that high manufacturing volume would be the key to cost. Perhaps six 'mega-producers' were to coalesce out of the 20 final assemblers in the Western world in an attempt to keep ahead in economies of scale. ... Finally, manufacturing would shift from the developed countries to the less developed countries as auto makers took advantage of lower wages to reduce manufacturing costs.
>
> (Altschuler *et al.*, 1986, p.181)

But there were considerable problems in moving towards **global products**. Market diversity and other differences (such as in product regulation for safety and other purposes) create a series of difficulties. An attempt by Ford to move its Escort model in this direction achieved only six common components between the U.S. and U.K. versions (Kaplinsky, 1988). Most important perhaps, as Altschuler *et al.* found in their car study, there is little evident demand for a globally standardised product. More generally, as you have seen in the earlier Blocks and in the earlier sections of this Block, there has been a retreat from standardised products. Increasing priority is attached by producers and by some consumers to higher and more consistent levels of quality, a greater range of functions within products, more product options, and greater product variety. As you will recognise from earlier in this Block, this is a shift towards **design-led development**.

There are two underlying sets of factors for this shift, both of which emphasise the interdependence of product innovation and design with manufacturing systems innovation.

The first is the development of production systems, primarily in Japan, which, largely through innovation in the organisation of manufacturing, found a territory between the low-cost, standardised ground of mass production, and low-volume, high-cost, craft production (see Block 5).

The Japanese-type systems were able to simultaneously process a variety of products, in low batch sizes at cost levels which matched those of mass production systems, and with consistent, far superior, standards of product quality. These production systems have evolved into new methodologies of manufacturing which are being adopted in the other industrialised countries. They can offer considerably greater freedom for product designers, both in the potential *scope* of design changes and in the speed with which they can be accomplished.

The second set of factors derives from the application of micro-electronics to production equipment, to design, and to manufacturing planning and control systems. As in the example of integrated systems which you saw in Section 4, these are central in enabling a more general shift towards **flexible production systems**. However, keep in mind the points about 'routines' (above). Such approaches are not easy to adopt, and can involve painful adjustments in work organisation and in other respects.

The next section will examine some of the implications of this shift.

## 5.4 TOWARDS MANUFACTURING SPECIALISATION ...

Currently, the shift towards greater product diversity and **production flexibility** is located in patterns of development in products and in manufacturing processes which appear set to last well into the twenty-first century. An important question is: does this indicates a shift away from the dominance of very large manufacturing and other organisations in product development?

There is no certain answer to this, for the outcome is almost certainly going to be mixed in character. First, the shift to flexible production is taking place within very large as well as in smaller companies, albeit with somewhat mixed results. This is partly because some of the re-learning processes that are involved are very complex, and because the changes marginalise much established investment and expertise (see below). Second, there appears to be a move towards what Rothwell and Gardiner (1989) term **design families**. They suggest that, in the process of design development, some companies have striven to identify and develop **robust designs**. These are designs with a deliberately designed-in element of *adaptability* or *'stretch'* which allows the development of a family of variants which will simultaneously meet the needs of a range of market segments.

In addition to such *parallel* stretch, robust designs also provide for subsequent, *sequential* stretch in later generations of the basic design. Some of the underlying ideas involved here are illustrated in Figure 47. Rothwell and Gardiner contrast 'robustness' with **lean design** – where there is limited potential for either parallel or sequential design changes. As an example of this, they compare the original (robust) design for the Boeing 747, which deliberately incorporated an element of stretch, with that of another airliner, the Lockheed L-1011. As can be seen in Figure 47, the 'stretch' in the 747 design provided the basis for a successful 'design family'. The L-1011 design, on the other hand, led on to only two design variants. Similarly, the European Airbus A340/A330 series was launched as a single programme which capitalised on *commonality* in components (an identical wing, cockpit and tail unit and the same basic fuselage) to create aircraft for different markets.

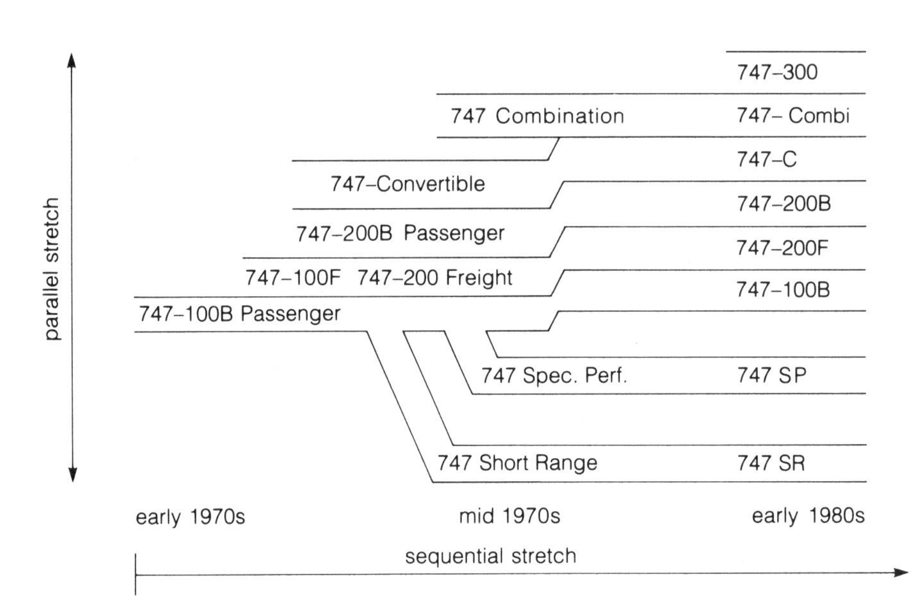


**FIGURE 47**
'DESIGN FAMILY' FOR THE BOEING 747 AIRLINER

While Rothwell and Gardiner's examples are primarily drawn from the specialised sectors of airframe and aero-engine manufacture, and from car production, the underlying concept has much wider applicability. For instance, it is evident in the extensive family of personal cassette players shown in Figure 48, which evolved from the original Sony 'Walkman', and it is used in clothing design, as you saw earlier. But the shift towards design families and other routes to product diversity is also associated with a transition to more rapid changes in product lines, and with reducing the lead times for product development – as you saw in Blocks 2 and 5. To develop production facilities which can process simultaneously a varying product mix at the necessary levels of efficiency, and which can rapidly assimilate new designs, generally requires high levels of investment in equipment, in training, and in other respects. Thus, like mass production, this is a product and production strategy in which the cost structure and the nature of the support infrastructure for manufacture and for sales often continue to favour the large company.

On the other hand, the problems of adapting to change across a wide span of technologies, the risk levels involved, and the levels of investment required have, in many cases, proved to be too extensive for even the largest companies to resolve alone. The growing complexity of product and production technologies, and the need to be close to the 'leading edge' knowledge and applications in specific technologies, have contributed to substantial increases in development and related costs. For example, Rosegger (1991) suggests that the cost of developing a completely new car model and of putting it into production can be around $3 billion. In this context, even the very large firms have had to move away from **integrated production** (in which one firm undertakes all the design and most of the production stages from raw materials processing to final assembly).

Integrated production is being displaced by a growing variety of approaches which sub-divide production methods by decentralising or sub-contracting elements of manufacture to specialist manufacturers – moving towards the elaborate type of supply chain which you saw in Section 4 of this Block. In many cases, this is leading towards a 'Japanese type' pattern of *inter-firm co-operation* in development and design, and of *risk sharing* between firms (where development costs and profits are shared between collaborators). In part, as Figure 49 shows,

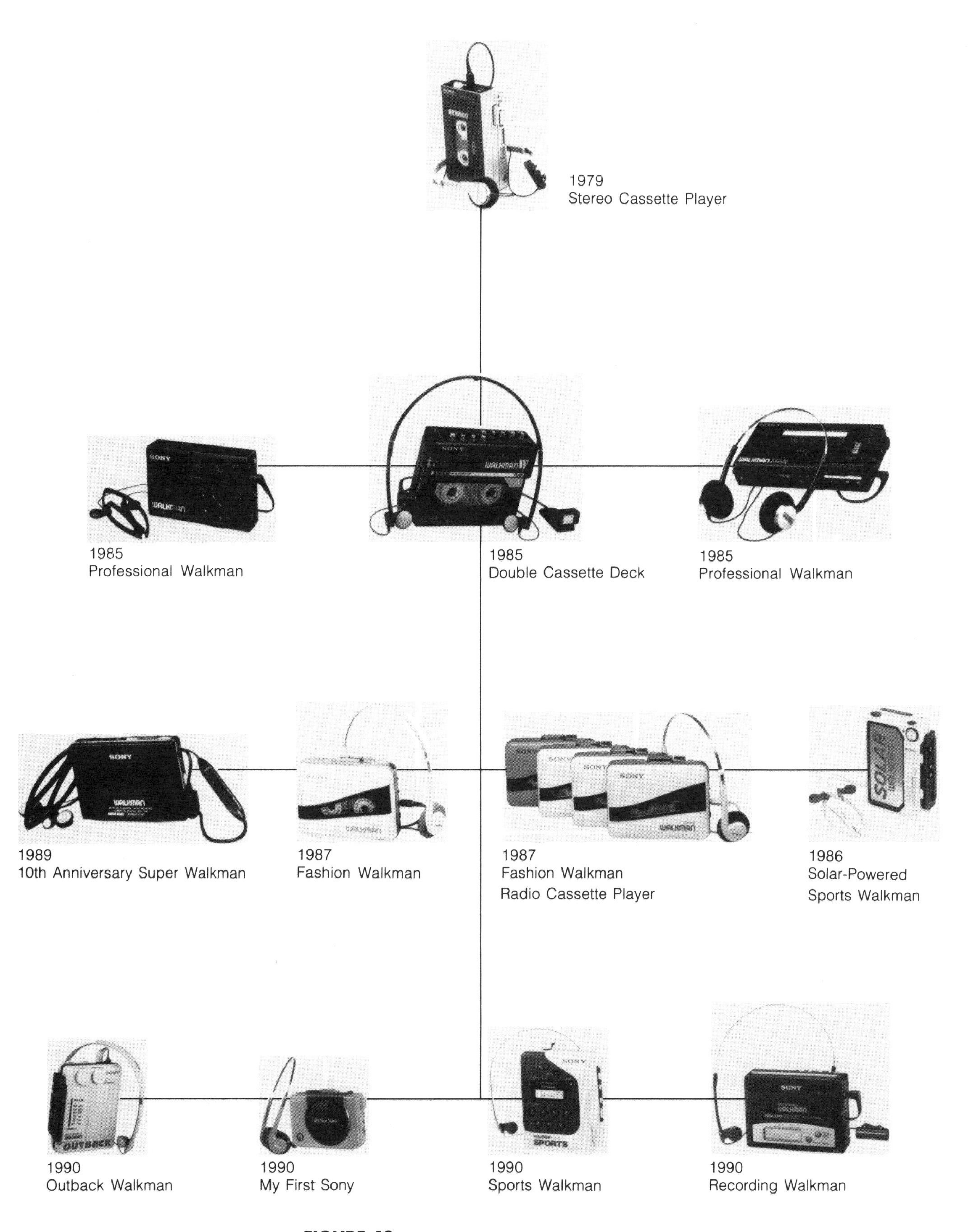


**FIGURE 48**
EVOLVING DESIGN FAMILY FROM INITIAL SONY 'WALKMAN'

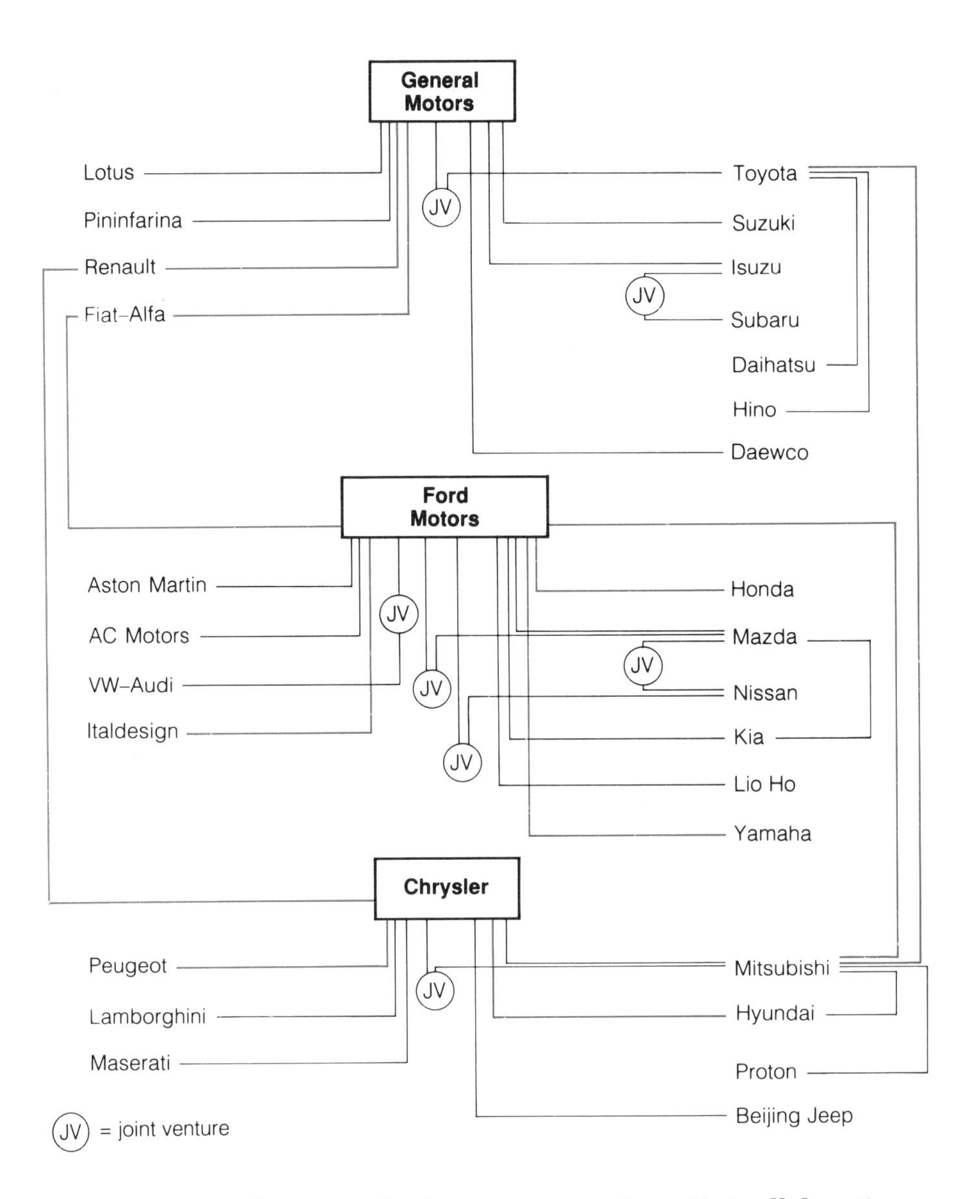


**FIGURE 49**
MAJOR CO-OPERATIVE RELATIONSHIPS OF U.S. CAR MANUFACTURERS

this is a matter of co-operation between competitors. But **collaborative ventures** also extend up the supply chain to include component and other suppliers. As is illustrated in Figure 50, this tends to be very extensive in Japanese firms and the success of the latter model is shifting the balance towards the collaborative pattern. This follows from the high levels of specialised competence which have been developed by the more competitive component, materials and other suppliers, and from their ability to operate more flexibly in production and in other respects.

## 5.5 ... AND DESIGN COLLABORATION

So, what appear to be the implications of the developments outlined above for the overall process of product development?

There are a number of implications. One is that the **diffusion** of technologies is speeded up, further quickening the pace of technological change. In addition, relationships between the design areas and other functional areas have come to be critically scrutinised. As you have seen, the design of even moderately complex products is not a self-contained activity, whether within a manufacturing organisation or in relation to the outside world. For instance, account has to be taken of the potential for cost containment by the use of bought-in components rather than those produced 'in-house'. But design teams do sometimes operate as

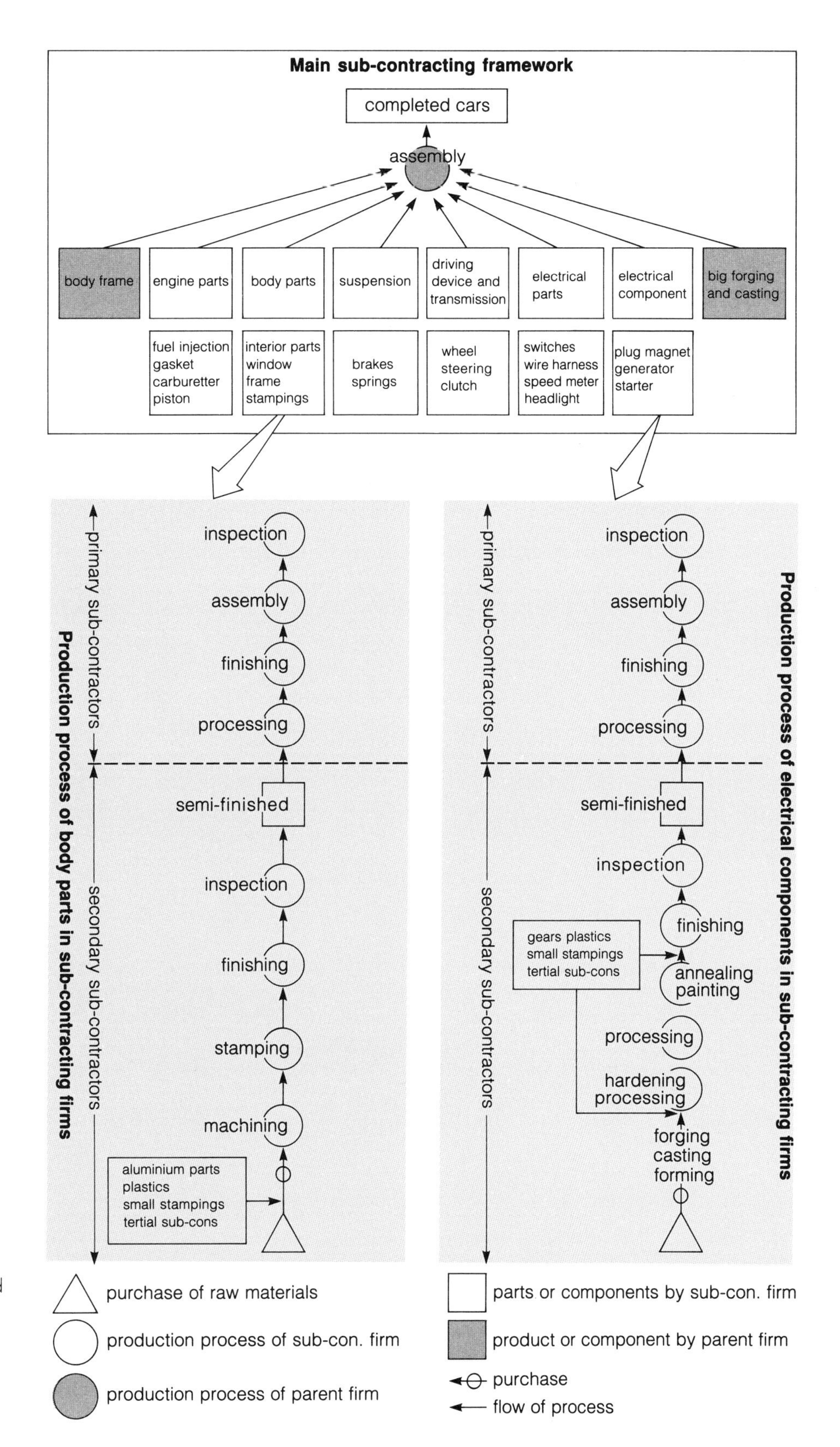


**FIGURE 50**

SUB-CONTRACTING – A JAPANESE EXAMPLE

Sub-contracting has been extensively used in Japanese car manufacture, and has tended to involve the design and manufacture of complete sub-systems by suppliers whose teams work in collaboration with the assembler.

though they were self-contained. Consequently, there has been widespread criticism of discontinuities between the design functions and other areas such as manufacturing. For instance, design teams sometimes do not give adequate consideration to questions of manufacturability so that significant modifications to designs may be necessary in the manufacturing start-up phase. These can be the source of substantial delays in start-up and of additional costs.

A further set of problems is parodied in the widely circulated cartoon included as the 'Postscript' in Block 2. In linear methods of development – or **serial engineering** – a project is handed on to successive functional areas. For instance, a new product design may require re-design of a part, or even all, of the manufacturing system involving, say, changes in the equipment used in assembly, and in plant layout. It may require the use of new skills for which shopfloor employees and supervision will need training. The inclusion of new types of part – such as electronic components – can mean that new suppliers have to be located by the purchasing department. Their capacity to meet the part specification at the required cost and volume levels and to provide quality assurance will need to be investigated. All of these activities add to the number of design modifications and to the total development timescale. Product development and modification accumulates as the specific concerns and constraints associated with each area are aggregated within the growing body of specificatory and other data. While some of the problems can be avoided by effective **iteration** (see Block 2, p.117), the iteration process carries a penalty. It can add further to the total development time as feedback from areas which are involved late in the development sequence provide feedback which leads to one or more re-design stages (as is depicted in Figure 51).

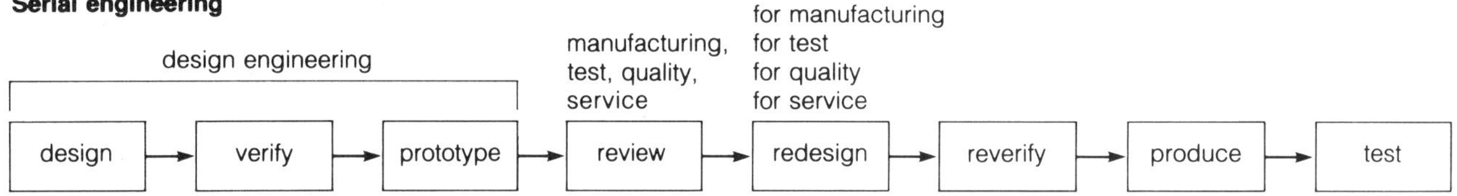


**FIGURE 51**
DESIGN-TO-MANUFACTURE ROUTE IN SERIAL (LINEAR) ENGINEERING

But it is possible to avoid much of the 'linearity' of the design process. This has been one factor in the product strategies of some Japanese manufacturers. For example, alternative methods of development contributed to lower development costs in Japanese car manufacturers and to development times which were about 50% of those of the U.S. car producers (Shina, 1991). In consequence, as part of strategies to restore competitiveness in American and European companies, there is pressure for continuous reduction of development lead times (as in the earlier example of '*quick response*' in Section 4, above).

Among other things, this pressure is leading towards use of **concurrent** or **simultaneous engineering** methods which were referred to in Block 2. In these, as is indicated in Figure 52, the objective is, so far as is possible, to undertake concurrently (rather than consecutively) all the activities required to develop, test, modify, and work up to standard manufacturing performance. But, to achieve this requires the direct involvement at a very early stage in the (modified) design cycle of all those concerned with one aspect or another of the total development–manufacturing cycle. It seems probable that there will be a widespread move towards the methods of concurrent engineering, and that this will contribute to some profound changes in design practice.

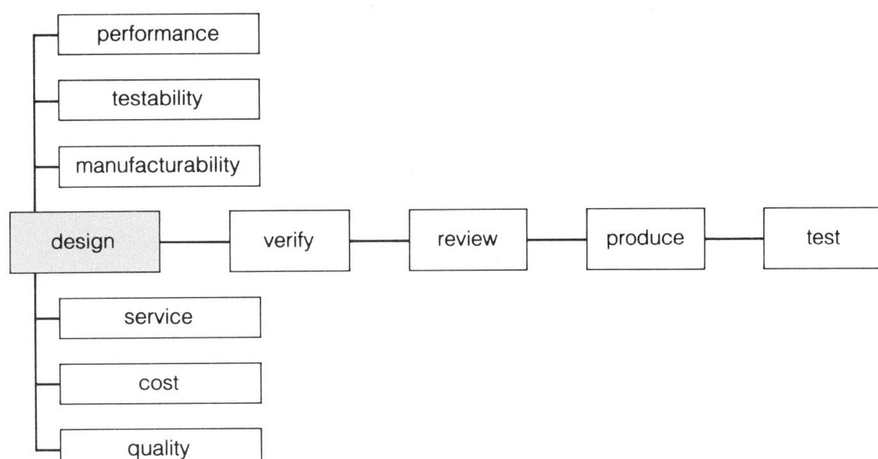


**FIGURE 52**
MODIFIED DESIGN CYCLE IN CONCURRENT ENGINEERING

The functional groups 'traditionally' concerned with design – such as development engineers and stylists – are increasingly likely to be involved in *collaborative* approaches to idea generation, concept development, and decision making. In Video 2, Section 2, you saw an example of this in a production review meeting at Electrolux which included representatives from production and marketing as well as from the design team. But changes in practice in a number of companies suggest that active interdepartmental collaboration is likely to be far more extensive. The trend is towards multi-disciplinary, multi-group team systems of various types, the membership of which may extend to areas such as purchasing and accounts. Furthermore, in the context of the general shift towards greater use of specialist suppliers and joint ventures, a growing number of design teams are likely to encompass members from outside organisations which, as you have seen, play an increasingly vital role in product development and competitive manufacturing.

> What types of problem might this present?
>
> ---
>
> See the discussion below.

The prospect of groups such as accountants who, in most companies, are not renowned for either their aesthetic or their engineering abilities, being involved in product development may appear to be an alarming one. But, of course, accountants already do have a central role in decision-making on design proposals. Concurrent engineering does not mean involving every interested person in every activity and decision in the various design stages. Depending on the scale of the project and the size of the firm, participation tends to be arranged through various working groups which are concerned with specialised aspects of the project. But this suggests a further set of problems – those of bureaucratisation through working groups, committees and so on.

In the ideal model, at least, one of the major objectives of concurrent engineering is to cut through the tangle of bureaucratic barriers and procedures which surround much current practice in 'serial engineering'. Broadening the base of the design activity is intended to provide an early dialogue – and consequent corrective action – on matters which might otherwise be confronted only at a late stage. The aim is to get new products to the market more quickly and without the delays which can follow from financial and other uncertainties which are uncovered only in the decision-making stage, and to avoid the teething troubles which may otherwise persist well into the production run. While it may be seen as involving 'non-designers' in some aspects of the design process, I have sought to show that, more generally, design is influenced by a range of other people and units across an organisation. It is, perhaps, preferable to make this influence overt, not least because the process of dialogue which becomes possible is a two-way process: designers are likely to gain insights into the reasons for some of the production and other constraints which may be encountered; but non-designers are likely to learn more about the many, complex problems encountered in the design process!

Overall, the types of changes that I have outlined – towards decentralised production in flexible systems, with a more varied, rapidly changing product mix, are now well established. In Section 5.2, I referred briefly to the way that changes in technology may be accompanied by developments in methods of organisation. The nature of current technological change is often interpreted as being part of a major 'generic' change, reaching across manufacturing in its effects. Yet, growing awareness of the significance of environmental constraints brings the duration of this paradigm shift into question. I will look at this next.

# 6 'GREEN' DESIGN

The proposition that design needs to take more account of the adverse effects of many products on the environment, and seek to minimise this impact, goes back to the work of Papanek (1972) – which you encountered in Block 1 – and before. The case has since been accepted to an increasing extent. From the mid-1980s there has been a marked trend towards the production and marketing of what are presented as 'environmentally friendly' or 'green' products. Likewise, many service sector products, from local authority systems for waste collection (for instance, bottle and can 'banks') through to insurance schemes ('ethical investment portfolios') have laid claim to be 'environmentally responsible'. This shift has taken place against a background of surveys which, in the industrialised countries as a whole, have indicated high levels of concern with a range of environmental issues. Indeed for a period in the late 1980s and early 1990s several surveys in the U.K. consistently indicated that concern with environmental issues was greater than for any other issue, including the health service, education and inflation (ENDS, 1990a).

However, it was evident that beneath such surface commitment and concern, there was a considerable element of confusion, lack of knowledge and contradiction. In a number of countries, governmental declarations of commitment to policies to reduce environmental damage have been matched by policies, such as large-scale road building, which have an overall negative environmental impact. Similarly, a number of companies marketed themselves (in terms of corporate image) and their products as 'environmentally friendly' on the basis that 'the environment is simply now becoming part of the branding process along with price, quality and other product benefits' (ENDS, 1990b).

In many cases, the substance behind environmental claims was limited. For example, the Consumers' Association (1990) was able to point to a variety of 'excessive', 'unexplained', 'meaningless' and 'unrealistic' examples of 'green' labelling on products such as paper goods, household cleaners, aerosols and batteries. One example was labelling a brand of washing-up liquid as 'phosphate free' when no washing-up liquid then sold in the U.K. contained phosphates. Some more extreme examples of deceptive green marketing – such as the advertisement by a major oil company that its lead-free petrol caused 'no pollution to the environment' – were exposed by the environmental pressure group Friends of the Earth in their 'Green Con' awards launched in 1989.

Against this background, it is not surprising that expressed levels of consumer concern contrasted with their actual patterns of purchasing activity. Various surveys found consumers deeply divided in their attitudes. For example, Ogilvy & Mather (1990) found distinctions between '*activist*' consumers who put environment before growth; *'realists'* who saw conflicts between profits and environment; *'complacent'* consumers who felt that someone else would solve the problem; and those who were *'alienated'* – overwhelmed by a sense of despair. Many people were uncertain about the meaning of the claims attached to green labels, and some regarded the issues as transient. There was scepticism about the motives of companies, some of which appeared to perceive the issue as offering a new area of niche marketing. Some manufacturers appeared to be 'simply jumping on the bandwagon, and using "green" labels as a marketing ploy' (Consumers' Assocn, 1990).

Given this degree of public uncertainty, how can real environmental considerations be incorporated into product design, and confusing or misleading claims be avoided? The first step is to define more clearly the criteria for assessing the **environmental impacts** of particular products or groups of products. The second is to establish a framework of controls on environmental claims by manufacturers. This was the basis of the European Community scheme for **environmental labelling ('eco-labelling')** introduced in 1992. Establishing such eco-labelling schemes quickly showed that the issues are more complicated than is often acknowledged. In general, although there is increased interest in the development of genuinely greener products on the part of many manufacturers, designers, retailers and consumers, it is often difficult to find solutions which are feasible, credible and generally acceptable, and the timescales involved can be very long. We will examine some of the basic problems, starting with an exercise to establish some of the underlying issues.

## 6.1 WHY 'GREEN' DESIGN?

Given that changes in product design, along with other more established measures such as pollution control, can play a vital part in tackling environmental problems, the starting point, obviously, is to find an initial working understanding of what we are looking at – namely, what constitutes 'green design'? It is also important to clarify what factors are associated with 'greenness', and why they may be important for design. So, I want you now to spend up to 20 minutes thinking about the following questions and jotting down some ideas. (In Section 6.4 you will be asked to consider how some specific types of product might be improved in the light of your ideas.)

### EXERCISE THINKING ABOUT GREEN DESIGN

**Question A**

Consider some of the products you use at home, at work, etc. What do you think might distinguish 'green' design from other approaches to product design?

(Spend about five minutes thinking about this.)

**Question B**

What factors appear to be contributing to an emphasis on 'green design' by manufacturers and retailers?

(Spend up to 15 minutes on this, jotting down some 'key points'. To develop a useful answer, you will need to think beyond general categories such as 'the environment'.)

**Do this exercise before going on to the next section.**

## 6.2 THE FACTORS BEHIND GREEN DESIGN

In this section I will provide brief answers to the questions in the above exercise.

### DISCUSSION OF QUESTION A

The distinguishing characteristic of a green approach to product design is that, in all the design stages (but most of all in developing the design brief, in formulating the product design specification, and in the choice of materials and manufacturing process), the potential environmental impact of the product is assessed with the specific objective of reducing this impact – compared with the preceding generation of products – and minimising it over the longer term.

Obviously, this begs the question of what constitutes 'impact'. One set of criteria, which defines the key areas of impact which an environmentally-aware designer should attempt to address, is shown in Table 4. We will consider their implications further in Section 6.3.

**Table 4 Criteria for green product design**

The environmentally-aware designer should aim to:

- increase efficiency in use of materials, energy and other resources;
- minimise damage or pollution from chosen materials;
- reduce to a minimum any long-term harm to the environment caused by use of the product;
- ensure that the planned life of the product is the most appropriate in environmental terms, and that the product functions efficiently for its full life;
- take full account of the effects of the end disposal of the product;
- ensure the packaging, instructions and overall appearance of the product encourage efficient and environment-friendly use;
- minimise nuisances such as noise or smell;
- analyse and minimise potential safety hazards.

*Source:* Burall, 1991, p.16

## DISCUSSION OF QUESTION B

The factors which lie behind the shift to green design are complex, and I can only remind you of some key points here. They include attitudinal and legislative changes (points (a) to (d) below) resulting from a growing awareness of, and response to, a variety of inter-related environmental problems (outlined in points 1 to 10).

### Attitudinal and legislative changes

(a) changing corporate perceptions of environmental issues and of the influence of these issues on consumer purchases;

(b) rising levels of voter/consumer awareness and concern about a range of environmental issues, and some apparent willingness to express this through modification of patterns of purchasing, and of waste generation and disposal;

(c) the influence of a growing number of designers who have embraced the view that, for social and moral reasons, and on practical grounds, product and other design needs to take account of environmental impact;

(d) a growing volume of increasingly stringent and complex environmentally-oriented legal requirements, standards and governmental regulations which have a mandatory impact on product design.

### Inter-related environmental problems

The original impetus for the above attitudinal and legislative changes arose from a variety of environmental problems and disasters that came to public attention in the late twentieth century: from the destructive effects of chlorofluorocarbons (CFCs) on the ozone layer and acid rain on German forests, to the nuclear accident at Chernobyl and the massive oil spill from the *Exxon Valdez*. Following many such disasters an understanding began to emerge that underlying these acute issues were a number of linked environmental problems:

1. the combination of rising world *population* levels with the extension of industrialisation across the world, and an associated rapid rise in global production levels;
2. the increasing scale and rate of depletion of *non-renewable physical resources* associated with rising industrial output;
3. the environmental impact of increased *exploitation* of supposedly *'renewable' materials* (for instance, where such materials are in practice not renewed – as in tropical rainforests);
4. the likelihood of major *climatic change* – i.e. global warming – which is increasingly accepted as resulting from emissions of carbon dioxide and other 'greenhouse' gases from some facets of industrial and agricultural activity and product use;
5. depletion of the stratospheric *ozone layer* resulting in increased levels of ultra-violet rays reaching the Earth's surface and increasing hazards to human and other animal life forms;
6. the highly skewed distribution of resource use between *developed and developing countries*, and related problems of poverty and distortions to the economies of developing countries associated with high levels of use of energy resources and materials in the industrialised countries;
7. rising levels of damage from *pollution* to the Earth's atmosphere, land surface, water resources (aquifers, rivers, oceans, etc.), for example from industrial emissions and the increasing use of agrochemicals;
8. fundamental damage to the *global ecology* as plant and animal species become extinct as a consequence of rising levels of industrial activity and intensive agricultural methods;
9. recognition of the pollution hazards associated with traditional methods of *waste disposal* – landfill and incineration; increasing costs and problems in the disposal of hazardous and all other 'waste' materials from production processes and from product use and disposal;
10. increases in the scale and types of hazards to human *health and safety* resulting from industrial, commercial and some leisure activities.

Some very large and complex issues are summarised in the last ten points – from changes to the global ecosystem, and the interactions of environment and economic development, to resource depletion and regional/local pollution. There is not space to discuss them here, but there are several excellent publications which do so – in particular I recommend *Our Common Future*, the report of the World Commission on Environment and Development (1987).

It is also important to recognise that the manufacture, use and disposal of different products are linked to the above environmental problems at different levels, from localised pollution through to world-wide effects such as global warming.

The magnitude and the urgency of the problems involved, at all levels, is probably best illustrated by the environmental impacts of motor vehicles, which I briefly discuss later in Section 6.4. But for now consider the following, seemingly mundane, aspect of motor vehicle use – disposal of vehicle tyres. In the U.K., an estimated 30 million tyres (about 400,000 tonnes) are discarded every year, adding to the already substantial accumulation of tyre waste. The accumulating 'tyre mountains' are dangerous in a number of ways, most obviously from fires which can burn out of control for long periods. For instance, a fire in a landfill site in Powys containing some ten million tyres which, after the initial outbreak in 1985, was still burning in 1992; a fire in a dump of seven and a half million tyres in Ontario burnt out of control for two weeks. These fires result in pollution of adjacent water systems and in land and atmospheric pollution which, in the Ontario example, included the release of heavy metals, phenols and PAH (polynuclear aromatic hydrocarbons, which are carcinogenic) into the atmosphere, affecting a wide area (ENDS, 1990c; 1992).

**SAQ 17**

You may remember the various creativity techniques in Block 3 which aim to stimulate you to generate new ideas for solving problems.

(a) If you had to generate ideas for solutions to the environmental problems posed by used vehicle tyres, what technique(s) might you apply?

(b) What new uses can you think of for discarded vehicle tyres?

(c) What ideas can you think of for reducing the number of tyres discarded?

The essential point that arises from the ten environmental problems listed above is that, while there are controversies about their causes and scale of effects, collectively they add up to a compelling array of reasons for changes in patterns of production and consumption, including attempts to design products that are less environmentally damaging. Most immediately, from the point of view of many companies, environmental considerations are resulting in new legal obligations and penalties, and the cost implications of design decisions – such as in materials selection and processing – are altering the decision-making framework. The key question then is, in what ways might changes have to be made? I consider this in the next section.

## 6.3 A 'CRADLE-TO-GRAVE' APPROACH

The previous section identified a number of general factors which add to the pressure for environmentally-driven approaches to product design. However, this is not only a matter of approaches to the design of future generations of products. Existing products also need to be evaluated, both to learn from and to remedy their shortcomings. Obviously, at a detailed level the nature of the tasks involved varies between products. But, in all cases, it is a matter of examining the *materials* from which they are made and the *energy* inputs that they require, and the *environmental impact* which results from processing materials and from energy use. This involves a number of departures from established practice, most fundamentally in approaches to the assessment of the pre- and post-manufacturing stages of a product.

As you saw in Sections 4 and 5, finished products are generally the summation of inputs from complex chains of materials and components supply. Locating the materials and components most appropriate for a product has tended to focus on factors directly related to financial and technical performance – such as the costs and availability of materials and the performance characteristics of a component. At the other end of the production process, manufacturers have tended to be concerned with a limited range of post-sale issues, such as cost and performance relative to competing products, warranty claims and after-sales support. These concerns have tended to be limited in duration, and have only occasionally extended across the full life of the product, and rarely into end-of-life disposal.

Thus, many of the impacts which a product may have on the environment are unlikely to be considered in the methods of product development which have tended to be used until now. It is increasingly recognised that a 'cradle-to-grave' approach needs to be used in assessing the environmental impact of a product. In other words, the choices open to designers need to be based upon an examination of environmental impact which considers *all* the life stages of a product. This 'total systems' approach is generally known as **cradle-to-grave assessment** or **life cycle analysis** (LCA).

The concept of product life cycles was introduced in Block 2, where Figure 4 illustrated the stages from manufacture and assembly, through installation, use and maintenance, to recycling.

> Look back at Figure 4 in Block 2 (p.21). What parts of the total life cycle of a product are missing from that diagram?
>
> ---
>
> See the discussion below.

Figure 4 in Block 2 provides a starting point, but it is necessary here to think beyond the simplified closed loop of that figure.

In part, this is a matter of extending the life cycle back to the 'cradle' – that is, to think back to the stages of raw material extraction and pre-processing into forms suitable for manufacture, since these are often the source of major, perhaps avoidable, environmental impacts. Most obviously, this is evident in the depletion of non-renewable resources, which might be reduced by greater use of recycled materials or materials from renewable sources. Where alternative materials or sources are available, they may need to be assessed and selected in terms of comparative environmental impact. For instance, in textile production, use of raw cotton or semi-finished fabric from countries with varying environmental regulations has to allow for the possibility of residuals from different types of pesticides and herbicides, not least because subsequent reprocessing may otherwise result in inadvertent water pollution and, potentially, prosecution.

At the other end of the life cycle, the environmental problems arising from the disposal of materials and components that are not recycled or re-used have to be considered. In the life stages between extraction and disposal, it is necessary to take account of any environmental impacts resulting from transporting materials, components and finished products from place to place, as well as from product use and maintenance.

In general, a satisfactory view of a product's environmental impact can be gained only by examining the range of inputs and outputs throughout the life cycle. A typical life cycle analysis chart is shown in Figure 53.

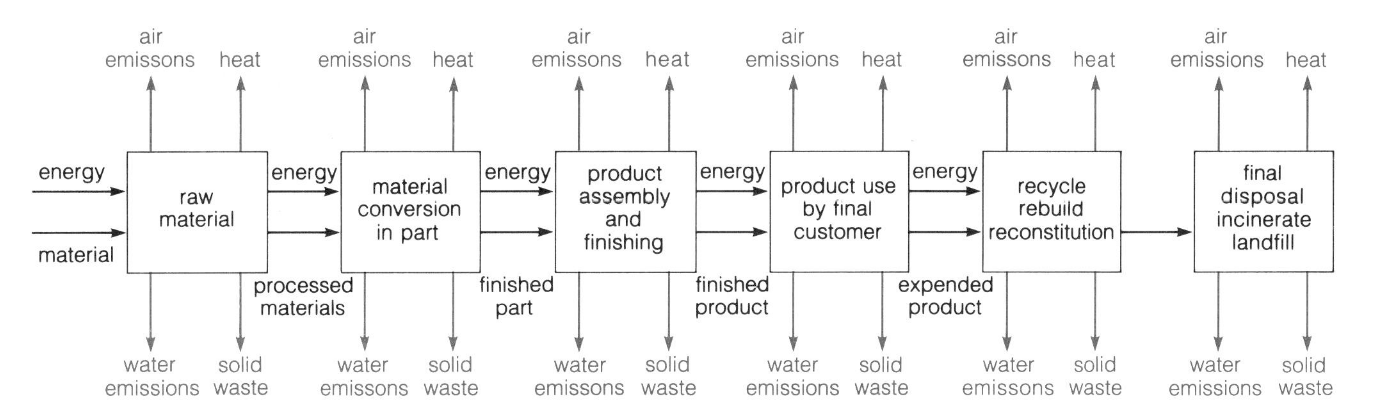


**FIGURE 53**

LIFE CYCLE ANALYSIS CHART

This relates the main life stages of a product to the inputs from, and outputs to, the environment.

What the figure seeks to emphasise is that:

- Use of energy and of materials continues through all stages of the life cycle for many products. Correspondingly, if issues of resource depletion are to be addressed, the processes of problem definition and solution in design have to be concerned with the full production cycle and with product performance in all stages.
- There are likely to be environmental impacts at all stages of the cycle – mainly air and water emissions and solid waste production, but also including outputs such as heat and noise. Many of these result directly from decisions made by designers, such as in the selection of materials and components.
- At present, only a small part of the materials discarded at the end of a product's life re-enter the production chain. Most materials at this stage are either incinerated or dumped in landfill sites or in the oceans. The extent to which recycling or reuse is possible in technical and economic terms is a direct function of decisions made in product design.

Figure 53 is, of course, simplified, and it understates the potential scale of the environmental impact of a product – for example, the manufacturing, installation and maintenance stages all generate large amounts of non-recycled materials, for instance in faulty and discarded components, and in packaging. Second, non-renewable materials constitute a large part of the discarded materials which end up as waste. Similarly, most of the energy used in the various life cycle stages is derived from non-renewable fossil fuels such as oil, gas and coal. Third, the interdependences between materials use and energy use need to be kept in mind – choices between materials can have different implications in terms of levels of energy use, whether in production, or in subsequent life stages. For instance, for many products, the choice of a high-cost, lightweight material, although increasing production costs, may reduce energy consumption over the full life of the product. In turn, consumers have to weigh an increase in initial costs against lower running costs in the 'use' stage of the cycle.

The importance of designing a product with its performance evaluated in terms of environmental impact across the total life cycle is emphasised by the way that use of energy and materials is distributed across the various stages. For example, it has been known for some time that in the case of a number of electrical and mechanical appliances such as refrigerators and cars, the larger part of the energy consumption occurs in *use* rather than in manufacture. Energy consumption for average European cars over a life of 100,000 miles is around ten times the energy consumption involved in production (Ferguson & Whiston, 1983).

However, it is only recently that detailed cradle-to-grave assessments of different products are beginning to reveal the full environmental impacts at the various stages. Notable among these is a study of the environmental impact of washing machines conducted in preparation for the EC eco-labelling scheme, mentioned earlier. (Other life cycle analyses for eco-labelling include ones for dishwashers, paper products, paints and insulating materials.) The washing machine study demonstrated very clearly that the 'use' phase can account for much the greater part of impact (PA Consulting Group, 1991). As is shown in Figure 54, over 90% of the environmental impacts in terms of energy use, water consumption, pollution and waste occur in use. This is the stage at which changes in product design (for example to increase energy efficiency) can have the greatest influence.

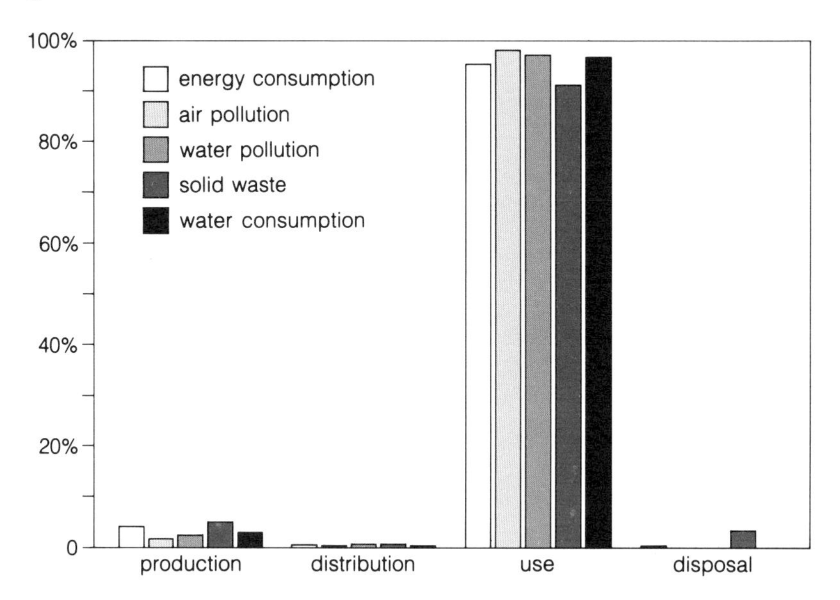


**FIGURE 54**
LIFE CYCLE ANALYSIS
Percentage contribution of life cycle stages to total environmental impact for washing machines.

Obviously, it is important that any cradle-to-grave assessment of environmental impact is undertaken in a systematic way. A typical framework (in this case the one used for the studies undertaken for the EC eco-labelling scheme) is set out in Figure 55.

It is also important to appreciate that life cycle analysis is a relatively new and complex technique and that the results of any analysis are dependent on the assumptions built in to each study and the data available. This has several implications. First, the conclusions of any particular analysis cannot be completely objective and are therefore open to challenge. Second, a full assessment of the environmental impacts of complex products, such as cars, with many hundreds of components and designs is very difficult. It is usual in such cases to concentrate on the most significant impacts. Third, it is also very difficult to make comparative assessments of the impacts of different technical alternatives for achieving the same practical tasks, for example comparing plastics and paper for packaging. In such comparative assessments it is necessary to make judgements about the relative importance of different environmental impacts – for example depletion of a non-renewable resource for plastics production against water pollution resulting from paper manufacture.

Life cycle analysis therefore is a systematic technique which provides vital environmental information to aid decisions on technical and design options, but it cannot substitute for human judgement and choice. How then, in practical terms, can environmental issues be addressed in the design of individual products? I take up this question in the next section.

| Assessment matrix | Product life cycle | | | | |
|---|---|---|---|---|---|
| **Environmental fields** | pre-production | production | distribution (including packaging) | utilisation | disposal |
| water relevance | | | | | |
| soil pollution and degradation | | | | | |
| air contamination | | | | | |
| noise | | | | | |
| consumption of energy | | | | | |
| consumption of natural resources | | | | | |
| effects on ecosystems | | | | | |

**FIGURE 55**
'INDICATIVE ASSESSMENT MATRIX' FOR CONDUCTING LIFE CYCLE ANALYSES
For each of the five product life cycle stages data is gathered on the impacts under seven environmental fields.

## 6.4 CONFRONTING THE PROBLEMS

I suggest that the issues need to be addressed on two levels. The first of these is ethical in character. As has been indicated earlier in this Block and in other parts of the course, some of the key driving forces in product development derive from market-related considerations of various types. In many cases, these appear to be unalloyed by any serious consideration of environmental or other effects. As we have seen, such one-dimensional orientations can nonetheless be projected in the garb of environmental concern, sometimes on spurious grounds. Perhaps as a reaction to this type of approach, much of the growing literature concerned with increasing the sensitivity of designers and other decision-makers to the range of environmental issues emphasise the need for a considered ethical base. For instance, Mackenzie (1991) argues that:

> The designer, as the principal determinant or creator of the product itself, has a direct influence on the amount of damage which will occur at each stage in the process. What materials will be used, and from where will these be obtained? How will the product be manufactured? Are particular processes required to give a specific effect or appearance? How will the product be used and disposed of – is it designed to be easy to repair, or to be thrown away? If it is to be disposed of, can parts of it be re-used or recycled? Designers, as creators or specifiers, are in a position to determine many of these issues.
>
> But designers also influence environmental impact directly, through their role as setters of styles and tastes. [...] Designers have participated fully in the disposable society, creating new styles with increasing frequency, and therefore necessarily building in

> obsolescence. They have often been criticised by environmentalists for failing to use their skills and influence to useful purpose. [...] Now, however, as individual values and business priorities are beginning to change, they have the opportunity to demonstrate that environmental considerations, along with social and ethical concerns, occupy a central position within mainstream design thinking.
>
> (Mackenzie, 1991, pp.11–12)

The second level of approach is that of evolving modified patterns of practice. What form might these take?

## SOME PRACTICAL EXAMPLES

We looked at some of the ways that practice might change in general terms in the preceding section. To take this further I want you to continue the exercise you began in Section 6.1 by considering some practical examples.

### EXERCISE THINKING ABOUT GREEN DESIGN (CONTINUED)

**Question C**

You have looked at a number of different products in the course. In the light of the criteria for green design and the environmental issues identified in Sections 6.2 and 6.3, and drawing on appropriate ideas from the course and your wider experience, suggest two or three ways in which products in the five categories below might be improved. The categories are:

- cars
- washing machines
- houses
- textile fibres used in clothing
- bicycles.

(Spend up to 20 minutes thinking about the question and jotting down some ideas.)

**Do this exercise before reading the discussion that follows.**

## CARS

Above all other products, the life cycle of the car (and other road vehicles) raises the widest and most complex range of environmental issues – from safety and the use of non-renewable fossil fuels to air pollution and climatic change. Added to this, automobile production is one of the world's largest manufacturing industries and an extremely powerful multinational political force. Because of this multiplicity of environmental impacts, combined with powerful economic and political interests, there are no easy 'green' solutions. Here, I have space only to identify some of the directions in which action might be taken.

### Vehicle manufacture

If you considered the full life cycle of cars, you may have first thought about materials use and energy use in raw material production and in manufacture. An obvious possibility is changing vehicle size and complexity – in essence, the larger and more complex the car, the greater the quantities of materials and energy that are likely to be used. In practice, the trend in Europe has been away from smaller vehicles, and you can probably think of the difficulties that lie in the path of any attempt to reverse this – or to otherwise constrain car design. A further possibility is to change the types of materials and manufacturing process used. But, whereas some changes (e.g use of water-based paints) produce environmental benefits, as you will see below, other changes are tending to increase rather than diminish the level of environmental impact.

### Emission reduction

A number of the ideas you may have thought about are likely to focus on propulsion systems and their associated inputs (fuels) and outputs (mainly gases of various types). Changes in engine design to allow use of lead-free petrol provided an elementary step towards reducing a particular pollution problem, and a further step has been the use of three-way catalytic converters to convert harmful exhaust emissions to other gases. Such 'cats' have been mandatory on all new cars in the U.S.A. and Japan since 1975 and have to be fitted on all new cars sold in the EC from 1993. But although cats are effective at reducing exhaust emissions of carbon monoxide, hydrocarbons and nitrogen oxides, they tend to increase engine fuel consumption and hence increase emissions of carbon dioxide, one of the main greenhouse gases linked with global warming. An alternative approach is to reduce emissions and increase fuel efficiency through the development of 'lean-burn' engines. But the EC's commitment to cats has – at least, for the time being – closed off this possibility.

Another approach you probably thought of would be improved fuel efficiency of cars through developments in engine technology and in lightweight, aerodynamic body design. The problem is that as vehicle efficiency has improved, people have tended to buy larger or more powerful cars rather than more economical ones. Thus, except in the U.S.A. where there has been a shift to smaller cars from the 'gas guzzlers' of the past, there has been little or no improvement in the average fuel consumption of cars in Europe over the last 20 years (Hughes, 1991). This is despite the fact that in a number of surveys, a large proportion of motorists have said they would buy a more economical car to reduce environmental impacts. This illustrates a more general problem. For many products there is a gap between consumers' 'conceived preferences' (what they say or think they *might* do) and their 'revealed preferences' (what they *actually* do). Designing cars that are highly fuel-efficient in no way guarantees their widespread sale. Green consumerism has its limits, and where a need for changes in consumption patterns is recognised, regulations or tax incentives are likely to be necessary accompaniments to green design.

Another idea you may have suggested is electric cars. Development of electric vehicles has been stimulated as a result of Californian legislation requiring the introduction by 1998 of a proportion of 'zero emission vehicles' (ZEVs) to address the serious smog problems of Los Angeles and other cities. Figure 56 shows one of many electric vehicle designs being developed by major manufacturers. It is important to note that this legislation concerns only one environmental impact – local air pollution – and the solution in no way involves a total life cycle analysis. Indeed, unless the batteries are charged from renewable energy sources, the use of electric cars will simply transfer emissions from car exhausts to fossil-fuelled power stations and may well lead to a net increase in overall air pollution and global warming effects. (The issue of oil resource depletion, which you may also have thought about, is more long-term.)

**FIGURE 56**
GENERAL MOTORS' 'IMPACT' PROTOTYPE ELECTRIC CAR

The Impact is powered using conventional lead acid batteries, but its lightweight glass-fibre body, high-pressure tyres and very low drag coefficient gives it an acceleration of 0–60 mph in 8 seconds, a top speed of 100 mph and a range of over 120 miles. The rationale for this performance is the need to compete with conventional petrol and diesel powered cars.

A further problem is that a shift to ZEV use, because it involves radically different technologies, requires the development of a new infrastructure for battery charging and for maintenance – which you may recognise from Section 5 in terms of 'complementary assets'. Moreover, market surveys indicate that unless the electric vehicles perform at least as well as conventional cars (or are significantly cheaper to own), it will be very difficult to persuade consumers to buy them (Power and Associates, 1991). Similar problems apply to attempts to tackle city air pollution in other ways, for instance, the use of alternative fuels such as hydrogen; and the development of 'hybrid' vehicles which combine internal combustion engines and electric propulsion – using the latter in inner city zones.

### Vehicle life

A further, more generally applicable, possibility you may have considered is that of extending the life of cars, not least because they are very high-cost items. But in practice the trend has been in the reverse direction – as their technological complexity has increased, car life times have generally decreased. For instance, since 1965 the median life for cars in the U.K. has fallen from about 15 years to around 10 years. Improved methods of body design (see Block 5) and parts standards would increase life spans. However, the results of such action would be mixed. Given that most of the environmental impact of cars occurs in use, a longer commitment to vehicles based on older technologies could outweigh any benefits of reduced consumption of materials.

**End-of-life**

Finally, you may have suggested ideas concerned with recycling, re-use and end-of-life disposal. Currently, this might appear to be one of the more positive instances of recycling. In western Europe as a whole, some 11 million cars are scrapped every year and an average of 75% of vehicle content is recycled (ENDS, 1991). But the level of recycling has tended to fall because of changes in the materials content of vehicles. In particular (as you saw in Block 5), the percentage of plastics used in European cars has risen from about 2% (by weight) to around 12% (Williams, 1991) and, for various reasons, these materials have generally not been recycled. Together with other non-metallic materials, such as glass and textiles, they are mostly disposed of in landfill sites. While the volume of waste from cars has increased, there has also been growing pressure on landfill capacity. Shortage of landfill sites, and more stringent environmental controls affecting waste disposal, have resulted in rising landfill charges. Without corrective action, it is forecast that the vehicle scrapping industry will become uneconomic, placing the whole recycling system in jeopardy (Williams, 1991).

Thus, at EC level, and particularly in Germany, there has been a growing intention to use legal regulation to sustain and to extend car recycling. In part, this has concerned the recovery system – by requiring car manufacturers to take back vehicles they have manufactured at the end of their lives, and seeking to ensure that a growing proportion of non-metallic components can be recovered for recycling. But the emerging regulatory regime is also focusing on the potential for reconditioning and re-using components – for instance, alternators, electric motors and gear boxes.

**SAQ 18**

The implications of actions to require vehicle dismantling, materials recycling and component re-use, if taken, are very considerable. From car bodies through to all components and the interior trim, car design will have to be undertaken in very different ways. From your experience and work on earlier parts of the course, what changes would you expect – in general terms?

At the time of writing (1992) the final characteristics of the German and EC proposals remained uncertain, as did their actual introduction. They have been criticised because the thinking behind them appears to be incomplete, not least because they are not based on a full life cycle analysis. For example, the much more significant issue of fuel economy is ignored – most notably in the case of large German cars and the consequences for fuel consumption of unrestricted speeds on German motorways. Also it is by no means certain, when all the energy and other inputs to the proposed recycling/recovery programme are taken into account, that the overall environmental impact will be beneficial.

You may have concluded from the above discussion that the best way of reducing the environmental problems posed by cars lies in shifting people to other forms of transport, rather than redesigning cars. This may turn out to be the case in the longer term. In the interim, some of these changes in vehicle design and technology may reduce the environmental impact of cars, provided consumers and the industry can be persuaded (or, if necessary, induced by legislation) to accept them.

## WASHING MACHINES

Making environmental improvements in the design of products such as washing machines is easier than for cars. You will recall that Figure 54 gave the results of a life cycle analysis of washing machines which illustrated the concentration of environmental impact in the *use* phase. This indicates the area in which most improvements can be made. You may have noted that resolving the problems is not the concern of washing machine manufacturers alone. It has also involved detergent manufacturers in changing the constituents of their products and their packaging (Figure 57), and in enabling a reduction in wash temperatures to reduce energy use. The replacement of mechanical control systems by electronic control and sensing systems has also contributed to reduced energy use, and to reduced levels of detergent and water use. The prospect of the introduction of eco-labelling to washing machines in 1993 stimulated many manufacturers to follow the lead of firms like Zanussi with greener designs (see Figure 58).

**FIGURE 57**
'GREENCARE' LIQUID DETERGENT

This product, produced for Sainsbury's, is formulated to help reduce water pollution. It also illustrates the potential for reduction of packaging by allowing refilling of the bottle from the plastic pouch. However, the benefits of such a system depend on whether the bottles are collected for recycling (or returned for refilling) and whether additional packaging is needed when transporting pouches rather than bottles.

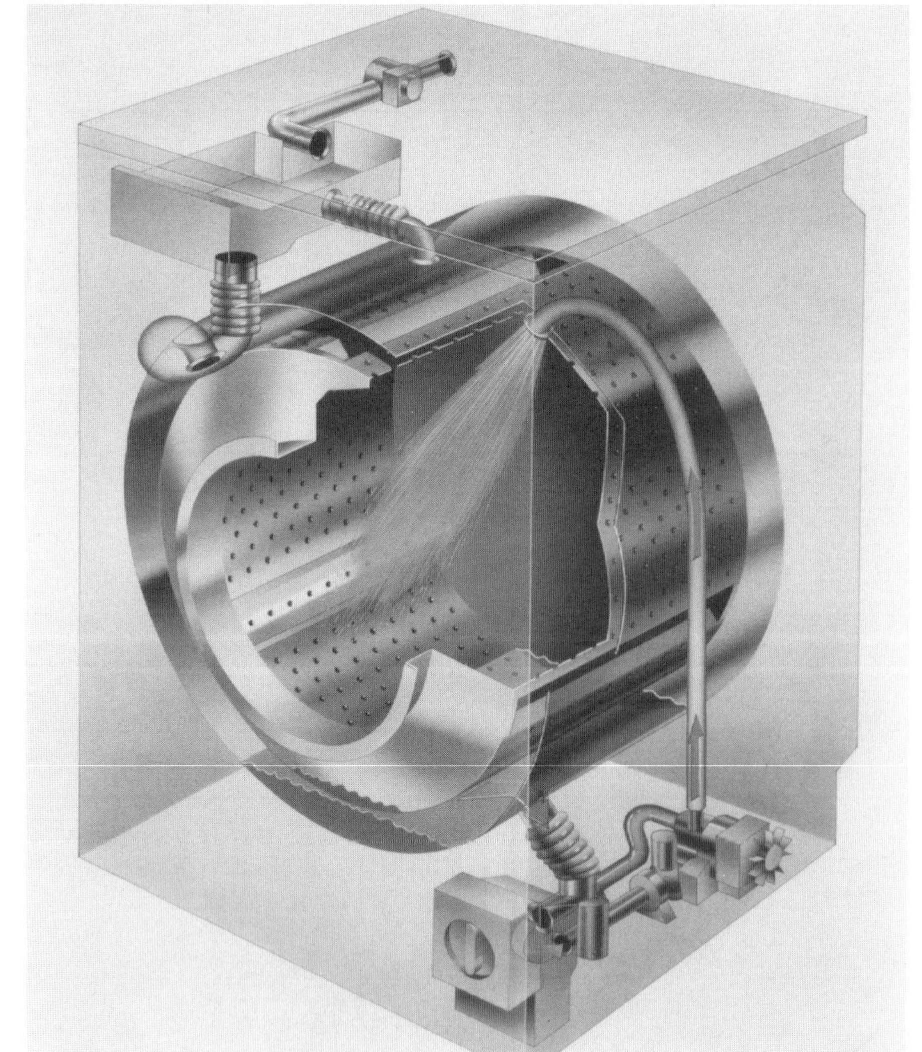

**FIGURE 58**
ZANUSSI JETSYSTEM WASHING MACHINE

The machine assesses fabric type, weight and volume of load to allow savings of water and energy. The water is circulated to the top of the machine and sprayed down on the clothes instead of soaking them in the drum.

## HOUSES

As with washing machines, the environmental impact of houses is very concentrated in the use phase, mainly arising from the energy consumed for water and space heating (or air-conditioning) and for lighting. You should recall from Block 4 that energy use in housing can be reduced by design to minimise exposed surfaces, orientation to make optimum use of solar gain and good insulation. But you will probably also know that energy consumption can be further reduced by measures such as energy-efficient heating (or cooling) systems and low-energy lights. You may also have thought of other devices, such as solar collectors for water or space heating, or mechanical ventilation and heat recovery systems.

Concern with environmental issues has stimulated a growing number of architects and developers to design and build low-energy houses, although the number in the U.K. (such as the designs in Figure 59) is still small compared with that in other countries.

(A)

(B)

**FIGURE 59**
LOW-ENERGY HOUSES

Two of a number of low-energy designs built in Milton Keynes in the 1980s and 1990s. Both designs are well insulated and are oriented for maximum passive solar gain. House (A) has a mechanical ventilation and heat recovery system, while (B) is similar to a group of houses with photovoltaic cells on the roof of the conservatory for generating electricity.

## TEXTILE FIBRES

This example was included to emphasise again the need for a careful investigation of environmental impact. If you looked back at Figure 41 (p.78) showing the components of the textile chain, you will have realised that environmental impacts arise at all the links from primary production to retailing, and beyond that to use and disposal.

In terms of choices based on environmental impact, it seems probable that synthetic fibres will often be a wiser choice than natural fibres – particularly cotton. Synthetics account for only a very small part of oil use – about 0.04%. In contrast, cotton, which takes up about 5% of the world's productive land, is dependent on large inputs of fertilisers, herbicides and pesticides – and may account for up to 25% of world use of the latter (Mackenzie, 1991). Also, synthetics are generally washed at lower temperatures than cotton and dry more easily, thus requiring less energy for washing and drying. (The relative environmental impacts of natural and synthetic fibres should become clearer when the life cycle analysis of textiles, to be conducted for the EC eco-labelling scheme, is available.)

## BICYCLES

As you will probably have decided, it is difficult to envisage significant improvements in bicycle design which go beyond those discussed in Block 3 – unless one includes current and future developments in materials technology. The bicycle provides the most energy-efficient means of personal transport for many circumstances – as can be seen in Figure 60 (which also shows the greater efficiency of most public transport). Furthermore, over the (generally long) product life, it has a very low environmental impact – primarily adding slightly to the tyre mountains. But it illustrates an important point made at the beginning of the course (in Block 1, Section 2) about the need to extend design *outside* the boundaries which are imposed by products considered in isolation. Ultimately, the problems associated with the use of bicycles (such as hazards from other vehicles), and those associated with other forms of transport, arise from the design – or rather the lack of design – of the overall transport system. To an increasing extent, product design will need to be located in ordered approaches to the design of large systems and infrastructures, of which

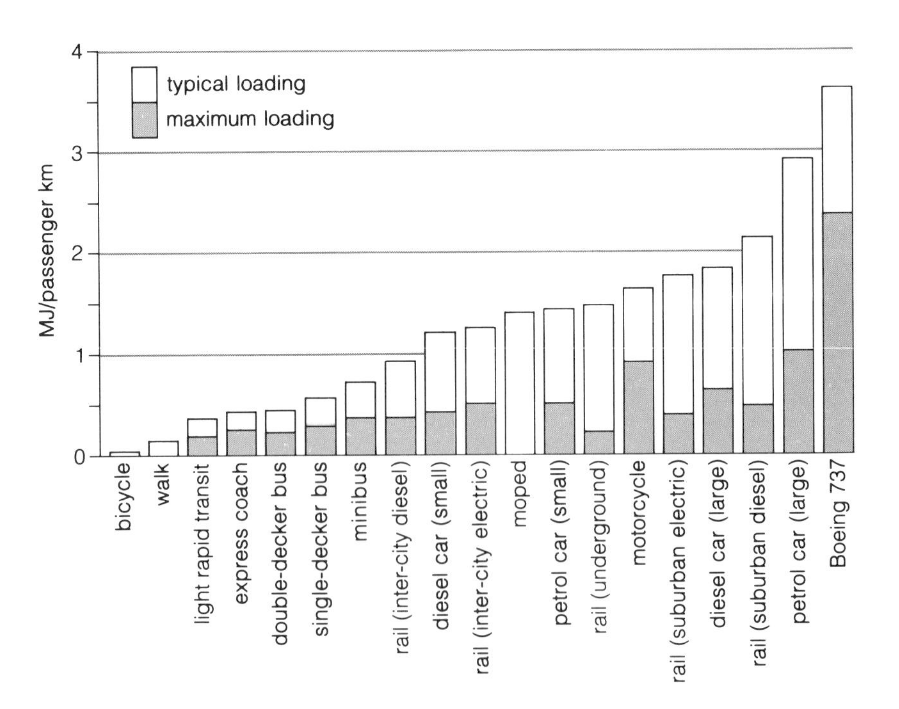


**FIGURE 60**
PRIMARY ENERGY REQUIREMENTS OF DIFFERENT MODES OF TRANSPORT IN THE U.K. (AFTER HUGHES, 1991)

(A)

(B)

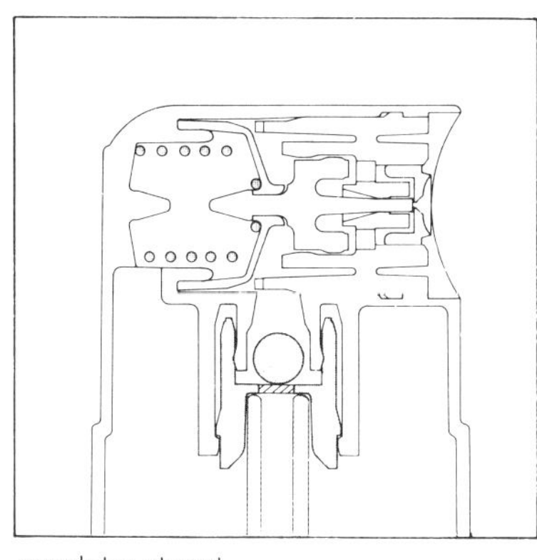


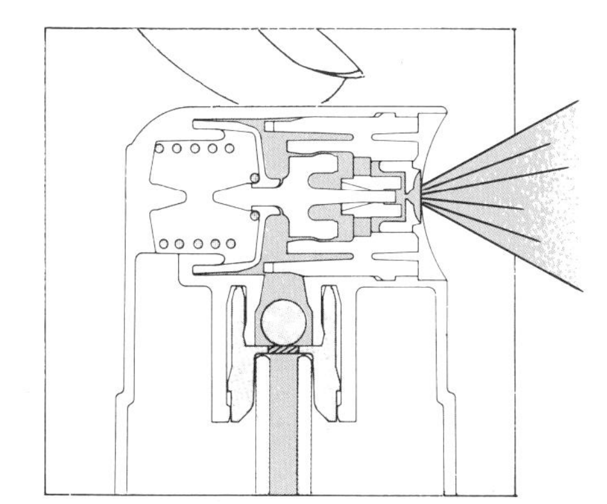


(C)

**FIGURE 61**

EXAMPLES OF GREEN – OR 'GREENER' – PRODUCT DESIGN

(A) Bench for outdoor use made from plastic bottles collected in a local authority recycling scheme. Other products made from mixed recycled plastics from the discarded bottles include compost bins and fencing.

(B) Shirt made from recycled flour bags by a cooperative in Zaire which provides employment for young women. The shirt is marketed in the U.K. by Oxfam.

(C) Atmosol aerosol system being developed by the British Technology Group eliminates both CFCs and hazardous liquefied gas propellants by the use of compressed gases such as nitrogen or air. The only difference from the traditional aerosol can is the regulator which replaces the button. When the regulator is depressed the valve at the top of the can opens and pressurised gas above the liquid forces the product out as a spray.

(D) Omni-Lite exterior floodlight designed for use with low-energy lamps. The company which developed it also produces domestic light fittings specifically for low-energy lamps.

(D)

## 6.5 END NOTE

As a final note for the end of this section, and one which also provides a fitting note for the end of the course, we return to the views of Victor Papanek:

> Design, if it is to be ecologically responsible and socially responsive, must be revolutionary and radical in the truest sense. It must dedicate itself to nature's principle of least effort, in other words, maximum diversity with minimum inventory [...] or doing the most with the least. That means consuming less, using things longer, and being frugal about recycling materials.
>
> (Papanek, 1985, p.346)

# ANSWERS TO SELF-ASSESSMENT QUESTIONS

### SAQ 1

The popular perception of clothing design is that it is essentially about styling and style changes. Styling is, of course, important – as is shown in Sections 3.5 and 3.6. But reference to structures and to the different properties of textile materials (i.e. the various fibre types) may have alerted you to the significance of more complex technological and other dimensions of design (see Section 2.7).

### SAQ 2

The pictures shown in the montage cover a long time span – from the 1920s to the current decade. Obviously, in choosing them, I had to be highly selective – so far as the archive material allowed. What I sought to do in this process of selection was to give some indication of the types of change that have taken place in the everyday clothes that are worn by large numbers of people – as opposed to the more specialised apparel that might be worn on special occasions and the clothing of the rich and of esoteric groups such as punks and rockers (see later illustrations). Apart from giving some indication of changing styles, they are also intended to give some hint of the link between apparel styles and changes in social attitudes and conditions which is discussed in Sections 2.5 and 2.6. Note both the changes in style across time, and the differences between members of different social classes – for instance, between Figures 13 (B) and (E); and between (D) and (F). Obviously, to produce successful designs, fabric designers and clothing designers have to be highly sensitive to such differences and sufficiently ahead of the underlying changes in values and attitudes – as well as possessing other knowledge and abilities, including awareness of the technical constraints and possibilities of manufacture.

The illustrations also give some hint of the importance of underlying technological advances – such as in types of textile fibres and in textile processing (see Section 4). This is most clearly indicated by the clothes of the company directors in Figure 13(A) – by modern standards, these appear to be ill-fitting, poorly finished and maintained. Also important in the changes indicated by the illustrations – though not shown by them – are changes in production technologies (in agriculture as well as in manufacturing) which have altered the relative costs of clothing, and contributed to changes in the volume of clothing demand (which we will look at later in this section).

### SAQ 3

The figure emphasises some of the compromises that have to be made in the overall process of design – between initial ideas and the 'shaping' effects of inputs from the market, from financial appraisal of designs, and from the actions of competitors. It also emphasises the iterative nature of the design process – but see the subsequent discussion of this in Section 5.

### SAQ 4

From what has been said so far, a lengthy development cycle follows from:

- design of products in ranges and high levels of variety within these ranges;
- the complexity of the production chain and associated technical constraints in production (to be considered further in Section 4);

- short product lives and thus frequent exposure to risk from uncertainties in consumer demand;
- the reflection of high risk levels in lost revenue from 'stock outs' and end-of-season price reductions;
- risk reduction which emphasises iteration and modification of designs.

**SAQ 5**

From the discussion up to this point, it may seem clear that market pull is a very strong factor – certainly, market conditions such as those discussed place emphasis on the product development strategies that emphasise rapid change. However, in the view of one of the T264 course team members (with which I agree), what it really demonstrates is a somewhat unique case of 'design push', in that the momentum of change is generated by a constant search for new design variants.

**SAQ 6**

Potential new entrants, it was suggested in Block 2, were one of four competitive dimensions which needed to be considered in external analysis. (What were the other three?) New entrants may bring 'new' goods or services, new technologies or novel applications of technologies which may force established companies to adapt their product range.

**SAQ 7**

Block 1 discusses psychological functions at a number of points. These are brought together in the exercise in Section 20 and in Figure 54 as three sets of motivations:

- encapsulating or expressing a public mood;
- expressing or symbolising a specific philosophy or world-view;
- making a public statement.

**SAQ 8**

As was suggested at the end of Section 2.6, there are some interesting similarities in developments in the bicycle, car and clothing industries. In all cases, competitive sport has been a stimulus to technological advances and/or their application in the development of new variants or types of product. Sport in this respect can be a powerful primary generator – although it is difficult to be sure whether this is *solution-focused* (i.e. to improve the chances of winning) or *problem-focused* (what is it that stands in the way of improved performance?).

**SAQ 9**

Block 2 describes the detail design tasks as involving decisions on a very large number of points of detail, as well as detailed drawing, testing and calculation, and of reference back to prior stages. As will be seen, this matches what is involved in clothing design. The process of reconciling design decisions with manufacturing considerations and the more specific financial constraints that are associated with manufacture are particularly important.

**SAQ 10**

Block 5, Section 3, identifies the knowledge base of engineers concerned with any given product as comprising the body of knowledge – primarily that related to materials properties and behaviour and manufacturing – which a combined group of engineers is able to draw upon. This may be from their own resources or it may be a matter of knowing where to look for particular knowledge and for information on how best to apply it for specific processes and purposes.

The significance of the knowledge base lies in its potential to underpin the success of particular designs and projects. But a well established knowledge base is a potential source of conservatism in engineering design. There can be an aversion to radical change which challenges the established body of knowledge. In such cases, an emphasis on the cautious extension of designs from the established body of practice can be fatal – for instance, where new opportunities are left unexplored – except by competitors.

**SAQ 11**

Three examples which come quickly to mind are:

- improvement in film as a product in response to the emergence of video;
- the development by the metal can industry of two-piece lightweight aluminium containers, in response to the emergence of plastic (PET) bottles for soft drinks;
- developments in high-speed rail transport technology (i.e. rail travel as a product) in continental Europe and Japan as a response to domestic air travel and road transport.

**SAQ 12**

Examples include:

- the increasing use of (computer-aided) simulation and modelling (Block 4);
- the introduction of 'lean production' in the car industry (Block 5);
- the shift to 'quick response' systems in the textile industry (Section 4, this Block).

**SAQ 13**

Although most of Block 3 was concerned with the role of individuals in creativity and innovation, Section 8 included a specific category of 'organisational barriers'. These included resistance to new ideas, particularly where these appeared to involve disturbance within an organisation, such as in threatening established skills or ways of working. But many of the other barriers to innovation and diffusion which were referred to may also be organisational in origin – such as financial barriers (discussed further in Section 5 of this Block) – and 'trialability'. However, it is important *not* to view 'organisational factors' as being essentially negative in character. Many manufacturing organisations have the capacity to focus considerable resources on the investigation and resolution of particular design problems – such as those classified as technical barriers in Block 3. Remember also the examples of organisational initiatives in Singer and in Xerox in overcoming the barrier of trialability.

**SAQ 14**

Under products which failed or which had difficulty in finding a market, you might have included the Sinclair C5 and the Cyclone vacuum cleaner.

An innovative product which was successful is the Sony Walkman.

The Sinclair C5 was probably not suited to the market it was aimed at, while the Cyclone vacuum cleaner encountered resistance from existing manufacturers. In contrast, Sony were able to generate a massive new market for personal cassette players, and to meet this demand from their substantial production resources and distribution system.

You might also have mentioned innovations like the Workmate® home workbench and the pneumatic tyre which eventually succeeded in overcoming resistance on the part of manufacturers and consumers.

**SAQ 15**

The most obvious 'processes and concerns' are those of marketing and production. Others include sales, product support/service, and the finance division. But remember that these are general departmental labels, and within them there are many other, smaller sections or groups – for instance, production engineering and the purchasing department, all of which have direct concerns with the development and the outcome of product modifications and of new projects.

**SAQ 16**

Earlier parts of the course have identified a diversity of prerequisites. Depending on the complexity of the design and the extent to which it departs from any existing design, the requirements are likely to include, that:

- adequate finance for development, launching and marketing is available;
- there is sufficient assurance from a suitably rigorous financial appraisal that an acceptable return on investment will be generated;
- there is good reason for confidence that the forecast market for the product will materialise at the forecast price levels;
- the design, including the proposed materials, bought-in components etc., is technically sound;
- the relevant safety and health regulations (and other market-related, non-statutory requirements) have been identified and can be met;
- there is adequate provision for pre-launch testing of product quality and reliability, and of manufacturing support in these respects;
- the product is manufacturable within the projected parameters of cost, quality and volume (this also has implications for the purchasing and supply of materials and components – i.e. requiring the close involvement of the purchasing division);
- adequate means exist for the distribution and sale of the product;
- post-sales support is available (parts supply, maintenance, operational support).

While the question asked you to identify four prerequisites, greater diversity is identified above. Some concerns are primarily technical in character, some financial, and so on. It would be surprising if views of the priorities and difficulties involved in relation to each of these concerns were generally shared. Hence discussion and reconciliation of concerns and objectives is likely to be important.

## SAQ 17

(a) The most suitable techniques are: brainstorming, brainwriting, checklists and analogies.

(b) Finding alternative uses for old vehicle tyres presents a number of problems because of their size, weight and dirtiness – which is one reason why there are so many lying around in dumps! Some tyres are given new treads and re-used, and it might be possible to devise ways of doing this more effectively. However, you might have thought of a number of alternative uses for them, including:

- safer swings for children;
- fenders for boats;
- door stops;
- soles for low-cost sandals and materials for shoe repairs;
- toys for large animals in zoos;
- low-cost protective edges in doorways and in places prone to damage by vehicles;
- as stabilisers to hold stacks of straw and other agricultural constructions.

These alternatives can only absorb a very small proportion of the available tyres. Two more promising solutions – in terms of the volume disposed of – have been found. One is reprocessing the rubber to provide a base material (rubber crumb) which can be used in the manufacture of a number of basic products such as traffic cones and safe play surfaces for children. This also yields a substantial amount of scrap steel. Mixed with aggregate, shredded tyres can also provide a road surfacing material. The second is to recover the energy stored in the tyres by burning them in purpose-built generating plants, fitted with suitable pollution control equipment.

(c) Possibilities are: to increase tyre life with new materials or by improving road surfaces; reduce vehicle use by various measures; reduce vehicle weight.

## SAQ 18

In the case of plastics materials, these will have to be selected – or rejected – according to their suitability for recycling. These and other materials will all have to be labelled in clear and durable ways to permit identification in the dismantling process. Car design will have to be undertaken with the subsequent ease of dismantling in mind – while also ensuring that vehicle integrity in terms of safety and other respects is maintained. Components or sub-assemblies which are appropriate for reconditioning and re-use will have to be designed with a corresponding robustness. Further, manufacturers will have to decide how they will establish suitable collection and dismantling facilities – a new part of their 'complementary assets'.

## ANSWER TO EXERCISE: A FABRIC CLAMP

The successful system involves the sealing of the upper surface of the fabric lay with an air-proof membrane enabling the use of a vacuum pump beneath the cutting table to clamp the fabric firmly in position. The reverse process – using an air cushion beneath the very heavy fabric lay (i.e. analogous to a hovercraft) – enables the lay to be moved easily and avoids the further stresses in the fabric which pulling it unaided would induce.

# CHECKLIST OF OBJECTIVES

Having completed your study of this Block you should now be able to do the following.

## SECTION 1

1 Apply marketing and innovation concepts in discussing the design of textile products.

2 Outline the length of the process of development of textile technologies and associated knowledge bases and the relationship between these and changing product market conditions.

## SECTION 2

3 Explain and define what are 'textiles'.

4 Explain the breadth of products within the overall categories of 'textiles' and within general sub-categories, and outline the technical and other commonalities between these product groups.

5 Apply an appreciation of design skills from other product areas to the design of textile products.

6 Identify social, economic and technological influences on the design of product types over an historical period.

7 Compare and contrast the product development process for clothing design with that for consumer durables given in Block 2.

8 Outline the influence that retailers can have in product design, with special reference to clothing.

9 Discuss the various influences on mainstream clothing design, including the interaction with high fashion.

10 Discuss the role and significance of new entrants to an industry with reference to the clothing sector.

11 Use the example of the textile industry to explain the role of non-price factors in competition.

12 Explain how various factors can influence the length of the product development cycle, with special reference to clothing.

13 Explain how product life affects market risk in the clothing industry, and describe the sources of market intelligence used by manufacturers to reduce that risk.

14 Explain how 'psychological functions' and the personal motivations of consumers are embodied and reflected in product designs.

15 Explain the types and significances of practical functions in clothing design for specialised purposes and for fashion products.

## SECTION 3

16 Make informed analytical comments about the design and manufacture of a given, relatively simple, product.

17 Relate a general set of requirements in a product design specification to the specification of textile products.

18 Indicate some of the factors governing selection of materials for related types of product, and indicate some of the design implications of different choices.

19 Outline some of the factors which might need to be considered in the forecasting of long-term trends in design, and explain the importance of such forward thinking.

## SECTION 4

20 Explain how manufacturing considerations can influence the design of a product.

21 Discuss conflicts between design for appearance, function and ease of manufacture with reference to the example of clothing.

22 Relate the properties of textile fibres and structures to the performance required of garments for manufacture and use.

23 Appreciate the significance of objective measures of materials properties for design and manufacture.

24 Explain the relevance of geometrical considerations of packing and symmetry in the detail design of garments.

25 Explain the nature of the main activities involved in detail design in a given industry (clothing manufacture); explain the nature of the detail design role, and outline some of the associated problems.

26 Compare and contrast the competitive strategies of U.K. and German manufacturers with reference to the clothing industry.

27 Use the concept of a 'supply chain' to outline the various stages of textile production and the interactions between manufacturers and suppliers.

28 Outline some potential effects on the processes and products of design arising from technological innovations in materials, manufacturing methods and design techniques themselves.

29 Discuss some implications of the introduction of CAD and other electronic systems with reference to fabric and clothing design and manufacture.

30 Outline changes in the role of clothing and textile design in relation to the co-ordination and integration of production across the textile chain as a whole.

## SECTION 5

31 Explain how organizational factors within companies can affect product design and the options available to designers.

32 Explain the role of 'routines' in determining the direction of product design carried out within organisations.

33 Outline some of the pressures for and against the design and manufacture of standardised global products.

34 Outline some of the factors tending towards increased flexibility and the decentralisation of manufacture.

35 Explain the concepts of 'robust' and 'lean' designs, and the significance of 'design families'.

36 Outline the reasons for a possible shift of design team activity to 'collaborative' models and related broadening in team membership.

37 Outline the drawbacks of 'serial' engineering which have led to moves towards 'concurrent' or 'simultaneous' engineering in industry.

## SECTION 6

38 Identify distinguishing characteristics of a 'green' approach to product design.

39 Outline reasons for environmental issues becoming influences on product designs.

40 Explain the relevance of life cycle analysis techniques in the assessment of the environmental impact of a product design.

41 Suggest ways in which designs of products could be altered so as to reduce the adverse environmental impacts they may have.

42 Discuss the advantages and drawbacks of different approaches to reducing the environmental impacts of cars.

43 Appreciate the limitations of green design as a way of tackling global environmental problems.

# REFERENCES


Altschuler, A., Anderson, M., Jones, D., Roos, D. & Womack, J. (1986) *The Future of the Automobile: The report of MIT's International Automobile Program*, MIT Press, Cambridge, Mass.

Anson, R. & Simpson, P. (1988) *World Textile Trade and Production Trends*, Economist Intelligence Unit, Special Report No. 1108.

Belussi, F. (1987) *Benetton: Information Technology in Production and Distribution: A case study of the innovative potential of traditional sectors*, Science Policy Research Unit Occasional Papers, No. 25, University of Sussex.

Borsboom, T. (1991) 'The environment's influence on design', *Design Management Journal*, Vol. 2, No. 4, Fall, pp.42–47.

Burall, P. (1991) *Green Design*, Design Council.

Carr, H. & Latham, B. (1988) *The Technology of Clothing Manufacture*, BSP Professional Books, Oxford.

CSO (Central Statistical Office) *Family Expenditure Surveys*, annual, HMSO, various years.

Colchester, C. (1991) *The New Textiles: trends + traditions*, Thames and Hudson.

Coleridge, N. (1988) *The Fashion Conspiracy*, Mandarin, p.170.

Consumers' Association (1990) 'Green labelling', *Which?*, January, pp.10–11.

Davies, S. (1987) *The Man-Made Fibre Industry in Western Europe*, Economist Intelligence Unit, October.

Dicken, P. (1986) *Global Shift: Industrial change in a turbulent world*, Paul Chapman.

Douglas, M. & Isherwood, B. (1980) *The World of Goods*, Penguin.

DTI (1991) 'Ecolabelling of washing machines', *Environmental Labelling*, No. 2, Winter, Department of Trade and Industry.

ENDS (1990a) *Backlash Against Green Promotions Confirmed By Latest Research*, Environmental Data Services Ltd, Report 183, April.

ENDS (1990b) *Reassessing Brand Values For The Green Revolution*, Environmental Data Services Ltd, Report 180, January.

ENDS (1990c) *Government Gets Conflicting Advice On Options To Reduce Waste Tyre Mountains*, Environmental Data Services Ltd, Report 186, July.

ENDS (1991) *Integrated Policy Urged On Car Recycling*, Environmental Data Services Ltd, Report 198, July.

ENDS (1992) *Tyre Fires Keep the Heat on Disposal Crisis*, Environmental Data Services Ltd, Report 204, January.

Ewen, E. (1988) *All Consuming Images*, Basic Books, New York, p.77.

Ferguson, E.T. & Whiston, T. (1983) *Product Life and the Automobile: A policy perspective for the Netherlands*, TNO (Centre for Technology and Policy Studies), Apeldoorn.

Forty, A. (1986) *Objects of Desire: Design and society, 1750–1980*, Thames and Hudson.

Gohl, E.P.G. & Vilensky, L.D. (1983) *Textile Science*, Longman.

Hoffman, K. & Rush, H. (1985) *Microelectronics and Clothing: The impact of technological change on a global industry*, Science Policy Research Unit.

Howard, R. (1990) 'Values make the company: An interview with Robert Haas' [Chairman of Levi Strauss], *Harvard Business Review*, September–October.

Hughes, P. (1991) 'The role of passenger transport in $CO_2$ reduction strategies', *Energy Policy*, March, pp.149–60.

Hunter, N.A. (1990) *Quick Response in Apparel Manufacturing*, Textile Institute, Manchester.

Information for Industry Ltd (1992) *Environment Business: Eco-labelling Supplement*, London.

Kaplinsky, R. (1988) 'Restructuring the capitalist labour process: some lessons from the car industry', *Cambridge Journal of Economics*, Vol. 12.

Kawabata, S. (1980) *The Standardisation and Analysis of Hand Evaluation*, HESC, Textile Machinery Society of Japan, Osaka, second edition.

Leigh, R., North, D., Gough, J. & Sweet-Escott, K. (1984) 'Monitoring manufacturing employment change in London, 1976–1981', *Industrial Sector Studies*, Vol. 2, Middlesex Polytechnic.

Mackenzie, D. (1991) *Green Design: Design for the environment*, Laurence King.

Mumford, L. (1945) *City Development*, Harcourt, Brace & World, New York, quoted in Hounshell, D.A. (1984) *From the American System to Mass Production, 1800–1932*, Johns Hopkins University Press.

Nelson, R.R. & Winter, S.G. (1982) *An Evolutionary Theory of Economic Change*, Belknap.

Ogilvy & Mather (1990) *Survey of Consumer Views on Green Products*, Environmental Data Services Ltd, ENDS Report 183, April.

OTA (1987) *The U.S. Textile and Apparel Industry: A revolution in progress*, U.S. Department of Commerce, Office of Technology Assessment, Washington DC.

PA Consulting Group (1991) *Environmental Labelling of Washing Machines*, PA Consulting Group, Royston, Herts.

Papanek, V. (1972) *Design for the Real World: Human ecology and social change*, Thames and Hudson (second edition, 1985).

Piore, M.J. & Sabel, C.F. (1984) *The Second Industrial Divide*, Basic Books, New York.

Power, J.D. and Associates (1991) *Focus Group Interviews with Consumers on Electric Vehicles*, quoted in Morales, R. & Storper, M. (and Associates), *Prospects for Alternative Fuel Vehicle Use and Production in Southern California: Environmental quality and economic development*, Lewis Center for Regional Policy Studies, Working Paper No. 2, May.

Rosegger, G. (1991) 'Diffusion through interfirm co-operation', *Technological Forecasting and Social Change*, No. 39, 1991.

Rothwell, R. & Gardiner, J.P. (1989) 'The strategic management of re-innovation', *R&D Management*, Vol. 19.

Rudofsky, B. (1947) *Are Clothes Modern?*, Paul Theobald, Chicago.

Sato, Y. (1983) 'The subcontracting production (*Shitauke*) system in Japan', *Keio Business Review*, Vol. 21, Part 1, p.9.

Shina, S.G. (1991) 'New rule for world-class companies', *IEEE Spectrum*, July.

Simmonds, P. & Senker, P. (1989) *Making More of CAD*, Engineering Industry Training Board, Watford.

Steedman, H. & Wagner, K. (1989) 'Productivity, machinery and skills: clothing manufacture in Britain and Germany', *National Institute Economic Review*, No. 128, Issue 2, May, National Institute for Economic and Social Research.

Sørenson, T. (1990) *Modern Methods for Assessing Handle and Tailorability in Fabrics*, Dansk Beklædnings og Textil Institut, Taastrup.

Teece, D.J. (1986) 'Profiting from technological innovation: Implications for integration, collaboration, licensing and public policy', *Research Policy*, Vol. 15, pp.285–305.

Tse, K.K. (1985) *Marks & Spencer: Anatomy of Britain's most efficiently managed company*, Pergamon Press.

Tubbs, M.C. & Daniels, P.N. (eds) (1991) *Textile Terms and Definitions*, ninth edition, Textile Institute, Manchester.

Walter, C. (1992) 'The impact of computer graphics on clothing design', in Aldrich, W. (ed.) *CAD in Clothing and Textiles*, BSP Professional Books, Oxford.

Williams, J. (1991) *Disposal of Vehicles: Issues and actions*, Centre for Exploitation of Science and Technology, London.

Wilson, E. & Taylor, L. (1989) *Through the Looking Glass*, BBC Books, p.105.

Wiseman, P. & Pellier, F. (1989) 'Process invention and innovation in the chemical industry', *Technology Analysis and Strategic Management*, Vol. 1, No. 2.

World Commission on Environment and Development (1987) *Our Common Future*, Oxford University Press.

## ACKNOWLEDGEMENTS

Grateful acknowledgement is made to the following sources for permission to reproduce material in this block:

### Figures

Figure 1: Courtesy of the National Museum, Copenhagen; Figure 2: National Museum of Wales/Welsh Folk Museum, Cardiff; Figures 3, 13A,B,C,D,E,F & G, 21A & B, 45: Hulton-Deutsch Collection; Figure 9: Linear Composites Ltd; Figures 10, 12: Joseph, M.L. (1977), *Introductory Textile Science*, Holt, Rinehart and Winston plc; Figure 11: Wild, J.P. (1988), *Textiles in Archaelogy*, Shire Publications Ltd, © John Peter Wild; Figure 16: adapted from Hunter A. (1990), *Quick Response in Apparel Manufacturing: A Survey of the American Scene,* © The Textile Institute; Figure 19 (top left & right): Makiko Minagawa; Figure 19 (bottom left & right): Colchester C. (1991), *The New Textiles: Trends + Traditions*, Thames and Hudson Ltd; Figures 20, 22A & B, 44: Associated Press, London; Figure 24: John Sturrock/Network; Figure 25: Rudofsky, B. (1947), *Are Clothes Modern?,* Paul Theobald, Chicago; Figures 26, 28, 30A & B, 31: Smart, J.E.(1985), *Clothes For The Job*, HMSO/The Science Museum; Figure 27: © Allsport / G.Vandystadt; Figures 29, 38: Dürkopp und Adler AG; Figure 32: Du Pont de Nemours; Figures 35, 36, 37A & B: Courtesy of Jaeger Tailoring Ltd; Figure 40: adapted from Gohl, E.P.G. and Vilensky, L.D. (1980), *Textile Science*, Longman Cheshire Pty Ltd; Figure 42: adapted from Pellier, F. (1985), 'Patenting and inventive activity on the nylon 6 intermediate caprolactam', M.Sc. Dissertation, University of Manchester 1985, © F. Pellier; Figure 47: Gardiner, P. and Rothwell, R. (1985), 'Tough customers: good designs', *Design Studies,* Vol. 6, No. 1, January 1985, Butterworth-Heinemann; Figure 48: Reprinted by permission of *Harvard Business Review.* An exhibit from 'Products: Variations on a theme', HBR Photo File, Vol. 69, No. 1, January/February 1991. Copyright © 1991 by the President and Fellows of Harvard College; all rights reserved; Figure 49: Reprinted by permission of the publisher from 'Diffusion through interfirm co-operation', by G. Rosegger, *Technological Forecasting and Social Change*, Vol. 39, No. 1, p.95. Copyright 1991 by Elsevier Science Publishing Company, Inc; Figure 50: *White paper on Small and Medium Enterprise 1982*, Yoshio Sato, The Society of Business and Commerce, Keio University, Japan; Figures 51, 52: adapted from Turino, J. (1991), 'Making it work calls for input from everyone', *IEEE Spectrum*, July 1991, © 1991 IEEE; Figure 53: adapted from Borsboom, T. (1991), 'The environment's influence on design', *Design Management Journal*, Fall 1991, Design Management Institute; Figure 54: Department of Trade and Industry (1991), 'Ecolabelling of washing machines', *Environmental Labelling*, No. 2, Winter 1991; Figure 55: Courtesy of Information for Industry Ltd; Figure 56: Reproduced with permission by General Motors Corporation; Figure 58: Courtesy of Zanussi Ltd; Figure 60: Courtesy of Peter Hughes; Figure 61B: Clive Boden Studios/Oxfam Trading; Figure 61C: Courtesy of the British Technology Group Ltd; Figure 61D: Starlowe Energy Ltd.

### Tables

Table 2: adapted from Kawabata, S. (1980), *The Standardization and Analysis of Hand Evaluation*, 2nd edition, The Textile Machinery Society of Japan, Osaka; Table 3: Sørensen, T. (1990), *Modern Methods For Assessing Handle and Tailorability in Fabrics*, © Tormod Sørensen, Dansk Technologisk Institut; Table 4: Burall, P. (1991), *Green Design*, The Design Council, © Paul Burall 1991.

### Text

The author wishes to thank the following for their help in reading and commenting on early drafts: Patrick Arnold Baker, Alan Brown, Ruth Carter, Alisdair Knox, Frank Lomax, Angharad Thomas, Clive Walter and Jennie Whitwam.

## T264 DESIGN: PRINCIPLES AND PRACTICE

### COURSE TEAM

| | |
|---|---|
| **Academics** | Godfrey Boyle<br>Catherine Cooke<br>Nigel Cross<br>James Forster<br>Georgy Leslie<br>Ed Rhodes<br>Joe Rooney<br>Robin Roy<br>George Rzevski<br>Philip Steadman<br>David Walker |
| **External Assessor** | Michael Tovey (Coventry Polytechnic) |
| **Consultants** | Michael Baker<br>Stephen Brown<br>Barry Dagger<br>Susan Hart<br>Jim Platts |
| **BBC Producers** | Cameron Balbirnie<br>Ian Spratley<br>Bill Young |
| **Course Manager** | Ernie Taylor |
| **Editors** | Keith Cavanagh<br>Garry Hammond<br>Rodney Wilson |
| **Graphic Designer** | Rob Williams |
| **Graphic Artist** | Keith Howard |
| **Media Librarian** | Caryl Hunter-Brown |
| **Secretaries** | Carole Marshall<br>Margaret Barnes<br>Jennie Conlon<br>Pat Dendy |